T0133030

The Third Lens

The Third Lens
Metaphor and the Creation of Modern Cell Biology

Andrew S. Reynolds

The University of Chicago Press :: Chicago and London

PUBLICATION OF THIS BOOK HAS BEEN AIDED BY A GRANT FROM THE
BEVINGTON FUND.

The University of Chicago Press, Chicago 60637
The University of Chicago Press, Ltd., London
© 2018 by The University of Chicago
Published 2018
Printed in the United States of America

27 26 25 24 23 22 21 20 19 18   1 2 3 4 5

ISBN-13: 978-0-226-56312-1 (cloth)
ISBN-13: 978-0-226-56326-8 (paper)
ISBN-13: 978-0-226-56343-5 (e-book)
DOI: https://doi.org/10.7208/chicago/9780226563435.001.0001

Library of Congress Cataloging-in-Publication Data

Names: Reynolds, Andrew S., author.
Title: The third lens : metaphor and the creation of modern cell biology /
    Andrew S. Reynolds.
Description: Chicago ; London : The University of Chicago Press, 2018. |
    Includes bibliographical references and index.
Identifiers: LCCN 2017044566 | ISBN 9780226563121 (cloth : alk. paper) |
    ISBN 9780226563268 (pbk. : alk. paper) | ISBN 9780226563435
    (e-book)
Subjects: LCSH: Cytology. | Cells. | Metaphor. | Science—History.
Classification: LCC QH581.2 .R495 2018 | DDC 571.6—dc23
LC record available at https://lccn.loc.gov/2017044566

♾ This paper meets the requirements of ANSI/NISO Z39.48-1992
(Permanence of Paper).

For Kellie, Clara, and Alice

The expansion of a field is always associated with the development of new analogies and a new language; but perhaps the newness of the language is always and only relative.

LAURENCE PICKEN (1960), xxxvi

Contents

Introduction

According to one very popular conception—it is perhaps even the dominant one—science is supposed to provide us with an objectively true account of the world. If that is correct, then there can be no legitimate place in the final scientific account of things for nonliteral or metaphorical language. Metaphor, after all, involves the use of words typically ascribed to one type of thing to describe something else, and the result, generally speaking, is a statement that is literally false. For example, "That man is a snake in the grass." Metaphors, like similes, allow us to draw analogies between two things, but with the difference that where similes typically employ the terms "like" or "as" ("That man is *like* a snake in the grass"), metaphor invites us to regard them as being identical in some fashion ("That man *is* a snake in the grass"). It is because metaphor typically says something that is strictly speaking false that many writers have insisted that although metaphor may serve an initial heuristic function in the creative process of scientific discovery (acting really as a form of simile), it can play no valid cognitive role in the formulation of scientific explanation, understanding, or knowledge of how things actually work. In this book I argue that metaphors can and do play important roles in just these scientific activities, and I do this in part by means of philosophical analysis of scientific explanation, understanding, and knowledge, but also by illustrating as a matter of fact just how influential metaphors have been throughout the history of the cell theory and cell biology.

The cell theory is justly celebrated as one of modern science's greatest achievements, ranked alongside Darwin's theory of evolution by natural selection as one of the foundational principles of modern biological science (Mazzerello 1999, E15; Nurse 2003). A beautiful integration of analysis and synthesis, it provides both a powerful reductionist perspective and a unifying generalization about life in all its diverse forms. It can be stated very simply. As the authors of one of the most popular university biology textbooks put it: "In life's structural hierarchy, the cell is the smallest unit of organization that can perform all activities required for life. The so-called Cell Theory was first developed in the 1800s, based on the observations of many scientists. The theory states that all living organisms are made of cells, which are the basic unit of life. In fact, the actions of organisms are all based on the functioning of cells" (Urry et al. 2016, 6). But to say that the cell is the basic unit of life is to give a very abstract definition; to understand better the specific nature and properties of cells scientists have resorted to various more suggestive descriptions, and it is for this reason that the cell theory has been so reliant on the use of analogies and metaphors. The cell concept was from its inception based on metaphor and has continued to be shaped by metaphors throughout its 350-plus-year history. What does this mean for our understanding of science as the attempt to provide an objectively true account of the world? Is there a legitimate place for metaphor in science? If so, what is it? Are metaphors of merely heuristic value, as many have insisted, helpful to the creative process of dreaming up new ideas and hypotheses perhaps, but restricted to the *context of discovery*; or might they play an important cognitive or explanatory role within the *context of justification*, as some others argue? (e.g., Hesse 1966; Bradie 1998, 1999; Brown 2003; Ruse 2005).

It has become more accepted within recent years that metaphors in science are not just inconsequential window-dressing or *façons de parler* used by scientists to communicate difficult ideas to a popular audience of nonscientists; they can be integral to the formulation of a theory and constitute the core of action-guiding programs of research. They also incorporate implicit value judgments about the nature of the subject matter under investigation and the proper route to its scientific understanding. Haraway (1976 [2004]), Beldecos et al. (1988), Martin (1990; 1991), and Lewontin (1991) provided pioneering studies in this regard of the influence of metaphors in the modern sciences of the cell, development, and genetics. Metaphors do make a real difference to how science is done. Studies of the metaphors of the genetic "code" and of DNA as an "information"-bearing molecule (Keller 1995, 2000, 2002; Kay 2000) have shown how metaphors can exert a powerful influence on the direction and content of science, even if the ideas they promote are not all that clearly defined (e.g., talk of a genetic or developmental "program"). But more recent developments in cell and molecular biology reveal that much of the important operations and mechanisms of cellular activity occur *outside* of the nucleus (in eukaryotes of course). In particular, the events associated with cell-to-cell "communication" and intracellular "signaling," which take place "upstream" of gene expression, from the "reception" and "transduction" of chemical, electrical, or mechanical "signals" via membrane-bound "receptors" or membrane "channels" through the diverse and complex "pathways" that eventually lead into the cell nucleus have become central to attempts to understand cell development, behavior, physiology, and pathology. Infatuation with DNA as a "master" molecule that determines all biological events has given way to systems biology, with attention now returning to the cell as a whole and as a constituent and active member of a dynamic "network" of interactions within bodies, communities, and ecosystems. The relatively new and increasingly important field of cell signaling and cell-cell communication is a fascinating and metaphor-rich area of science so far underexplored by historians and philosophers of science.[1] Attention to it (see chapters 3, 4, and 5) teaches important lessons about the various roles metaphors play in current cutting-edge research in developmental biology, synthetic biology, cancer biology, and drug design.

In fact, since Robert Hooke first spoke of cells in 1665, the concept of the cell, and the cell theory that later developed from it, have been intimately shaped by a series of metaphors. To say that all living things are made of cells is one thing, but to say what these cells *are like* (their properties and behavior) is another. Because the science of cells concerns novel objects and events of a typically microscopic scale, scientists have been forced to use analogies with more familiar everyday phenomena to understand them and their relationship to the larger bodies of which they are a part; and since cells are supposed to represent the fundamental units of life, the cell theory raises deep conceptual issues about the part-whole relation between cell and organism. For these reasons biologists have relied quite heavily on metaphor and analogy to think about the nature of cells and to understand their causal relationship to one another and to the organism and body as a whole.

Metaphor is, as the book's title suggests, the "third lens" through which we see cells, the first two being the objective and ocular lenses of a compound microscope. This is, evidently, to use a metaphor to describe metaphor. Referring to metaphor, especially as it occurs in science, as a lens through which we see the world is a very common practice, and it will also be critically assessed in this book. The metaphor that metaphor is itself a *third lens* is represented by the image on the book cover. The image depicts the renowned cytologist E. B. Wilson (1856–1939), seen with pen and paper writing up the observations he has made through his microscope. Metaphorical language, I argue, has been essential not only to the activity of *describing* cells but also to *seeing* and *understanding* them, and has played no less a fundamental role than the literal and material lenses of the microscope.[2]

Chapter 1 begins with a brief discussion of metaphor and then provides a history of the key metaphors associated with the cell concept and the development of the cell theory from the seventeenth to the early twentieth century. Cells have historically been conceptualized in accordance with two fundamental categories of metaphor, or what the German philosopher Hans Blumenberg called "background" (*Hintergrund*) metaphors (Blumenberg 2010). The first category is of human *artifacts*: cells have been conceived as rooms or spaces enclosed behind a solid wall (the original cell concept), as building stones or blocks, chemical laboratories or factories, and as various types of *machine* (electronic computers being the current favorite). Artifacts are familiar to us because we create them, and as a consequence we understand well how they work. As a result, much of current thinking about cell function is dominated by the metaphor CELLS ARE MACHINES. The second background metaphor is from the category of *organisms*: beginning in the nineteenth century cells (including those of the human body) have been called elementary organisms (*Elementarorganismen* in German), essentially similar to unicellular amoebae and other protozoa, as living rich social lives as citizen-members of vast plant and animal bodies, described as "cell-states" or "societies of cells," in which they make decisions about which developmental "fates" to pursue (shall this stem cell become a neuron or an epidermal cell?). Talk of a cellular "division of labor" is still central to current discussions of the evolution of multicellular fungi, plants, and animals as one of the major transitions in the history of life. Hence the metaphor CELLS ARE (SOCIAL) ORGANISMS remains important for modern biology.

While some biologists invoke the organicist metaphor of the cell society to make sense of our compound anatomical constitution and its development (*e pluribus unum*), others—more concerned with the question "How do these little cell organisms *work?*"—adopt the rival philosophy of mechanism. Chapter 2 looks at the mechanistic metaphors of early cell physiologists and biochemists (from the mid-nineteenth into the first half of the twentieth century), who preferred to think of the cell in a more reductionist fashion as a chemical laboratory or factory. These metaphors highlighted better the physical processes that must be occurring within the boundary of the cell membrane to make life possible. By analogy the cell factory was assumed to be performing these tasks by means similar to those being perfected in the chemical laboratories of nineteenth-century scientists, without recourse to mystical vital forces. The continued success of this mechanistic approach is evident in current cell and molecular biology, where talk of cells consisting of complex "protein machines" is now the norm. Proponents of these reductionist and mechanistic metaphors (molecular and synthetic biologists, for instance) tend also to be practitioners of what has been called the "engineering ideal" in biology (Pauly 1987). These mechanist/artifact metaphors align naturally with the epistemic thesis that knowing is making (cf. Keller 2009a), that a thing is not properly understood until it can be constructed or replicated by humans at will. If we call a cell a machine, we naturally expect to identify its component parts and to figure out which are causing others to move and behave in various ways, etc. And once we have done that, we can begin to manipulate the system, to "improve" it to fulfill ends we deem desirable. By contrast, if we think of the cell as an organism, and a social one in particular, we naturally study its gross behavior, how it interacts with others, and the causal effects those interactions have on both it and its "neighbors." These two sets of metaphors function therefore somewhat like lenses of different magnifying power by focusing attention at different levels of biological organization.

Chapter 3 tells the story of how the development of tissue and cell culture technique in the early twentieth century, which allowed scientists to study animal cells as live individuals outside of the body, spurred further articulation of the social metaphor into an approach called "cell sociology." In developmental biology "cell sociology" regards a developing embryo composed of multiple cells as analogous to a society of interacting human agents whose behavior and future fates are the products of social interactions experienced by each cell as a member of a particular peer group of similar cells and its interactions with groups of dissimilar cells. Cells are not, according to this view, little automatons playing out predetermined developmental programs written in their genetic code, but complicated social beings whose collective behavior brings forth emergent properties (known as group or "community" effects) that are unattainable to single cells individually. In fact, a tissue cell from a mouse or human placed in isolation in a petri dish dedifferentiates to a more primitive state, before ultimately dying in a process referred to as cell "suicide" (also commonly known as apoptosis). This type of cell death, otherwise known as "programmed cell death," is under tight genetic control and forms an essential component of normal embryogenesis and tissue maintenance. As each cell in the body can be instructed at any time by its neighbors to engage this suicide program (note the mix of social and mechanistic metaphors), the organism as a whole is said to exert a form of social "control" over its constituent cells. Cancer, it is now believed, can arise when cells fail to respond to these death "signals" and begin to proliferate in a socially irresponsible fashion. To understand *how* cells go about killing themselves (committing suicide), scientists resort to mechanistic metaphors of programs and "signaling pathways" and "circuits" involving various molecular parts or modular components, in analogy with those found in modern computers and other electronic devices.

Chapter 4 details how classical cell theory has been significantly revised in the last few decades by the study of cell-to-cell communication. Far from being inert building stones, cells are in constant communication with one another and with their environment, and this profoundly influences their behavior. But while communication is an intrinsically social phenomenon, to understand the causal details of *how* a cell receives and processes a message from its milieu, molecular biologists rely upon mechanistic engineering metaphors. The science of cell signaling, now one of the most significant and promising areas of biological and biomedical research, is constructed upon the metaphors of "signal transduction," gene regulatory "switches," and others borrowed from electronic engineering and cybernetic theory. Medical researchers and synthetic biologists today seek to "reprogram" cells and "rewire" their communication circuits and "signaling pathways," in efforts to treat cancer and other diseases. Development of new drugs rests upon intervening in what are increasingly recognized to be complicated intracellular signaling "networks." Driving all these projects is a conception of cells as devices open to improvement through human engineering, and of course as potentially lucrative intellectual property subject to patent protection and commercial marketability.

Current understanding of cells then makes use of both social (organicist) and mechanist metaphors: the cell is a being with particular characteristics, which may change over time as a result of its history of social interactions with other cells, but it is also a machine that performs some kind of work or fulfills some function as a result of the properties and behavior of its component parts (e.g., proteins, DNA). These two very different sorts of metaphors facilitate distinct yet complementary kinds of investigation and understanding, as the following chapters will show. It is striking how scientists invoke both types of metaphors, sometimes even in the same sentence, in their attempts to describe and explain cell behavior and function.

Another chief object of this book is to provide an account of *why* metaphor is of value for science, but one that is self-reflexive and critical of the metaphors (or better, meta-metaphors) that seem to be used almost unconsciously in discussions of this question. This is the topic of chapter 5. The dominant account of metaphor's utility for science relies on what I call the "perspective" theory, according to which metaphors are said to provide a new or useful perspective or point of view from which to see some subject of investigation. Such talk is itself metaphorical, and given that viewing an object from one perspective precludes seeing it from others, we ought to wonder what aspects of metaphor's role in science gets "pushed to the periphery" of our "mental vision" when we choose to employ this particular metametaphor. The perspective theory is a natural one, given our great reliance on vision for learning about the world, and it is intimately associated with what John Dewey called the "spectator theory of knowledge," i.e., the thesis that knowing is a kind of internal mental vision based on passive reception of external sensory data. Despite the book's title, I argue that we should not always think of metaphor's role in science as providing a visual-like perspective or a passive lens on reality, when in some cases metaphors act more like tools that allow us to get a grasp or handle on some aspect of the world so that we can dissect it, identify the parts, put them back together, and in some cases redesign them so as to better suit our own specific ends, as when we re-engineer cells through genetic or other molecular intervention.

Thinking of metaphors as cognitive tools or instruments with which scientists investigate reality is better suited to a pragmatist understanding of knowing. Its implications for understanding the kind of knowledge science provides are significant, for just as there are different sorts of tools for different jobs, so different metaphors for the cell can assist with different sorts of tasks without any one metaphor being the ultimately correct one; and just as there is no universally correct or true tool, so there may be no universally correct or true account of how metaphors function in science that can be captured using just one of the standard metaphors for metaphor.

But *must* we use metaphor to talk and think about metaphor? Can we not describe how metaphors function in science in purely literal language? The problem is that metaphor is such an abstract thing that we have no immediate literal descriptions available with which to work; we are forced–if we wish to say much of interest about it–to employ analogies with other, more familiar, aspects of human experience, like viewing an object from different perspectives, manipulating it with various tools, or carrying something from one place to another (which is in fact the literal translation from the Greek *metaphora*, μεταφέρω). My conclusion, then, is that we must recognize the limitations of each of these metaphorical perspectives (or metametaphors) and employ them all to get the most complete understanding of how metaphor works in science, just as scientists employ many different metaphors to understand the various aspects of cells.

Chapter 5 also takes up the question whether metaphors can ever be truly explanatory. I argue that they can, because providing explanations is a pragmatic affair for which there is no universally valid formal model with literal truth as a requirement. I also consider the nature of so-called '"dead" metaphors and the process leading to *polysemy*, by which once-novel metaphors gradually lose their metaphorical status and are regarded as literal.

That the metaphorical-literal distinction is more fluid than normally recognized has implications for the question of scientific realism, the topic of chapter 6. What are the implications for our understanding of science that it relies so heavily on metaphor? Scientific realism is frequently understood to include the thesis that science aims at literally true theories. But discussions about the realism question tend to conflate the idea that science aims to provide a *literally* true account of the world with the quite different, and more philosophically ambitious, idea that it aims to provide an *objectively* true account of the world, that is a description of reality as it *really* is independent of us or *in its own terms*, as it were. Both theses share the assumption that science ought to result in one uniquely correct account of the world. But I argue we need not accept either of these proposals (literalism being too restrictive and objectivism unattainable), nor the assumption that science should aim for one uniquely correct account or description of the world (or even that each of the special sciences should aim for one uniquely true account of the limited subsystems of nature with which they deal individually). This is for two reasons: (1) science has several aims, true description being only one of them–explanation and intervention are two others, and for these tasks there may be no one best solution or account, nor need it be free of metaphor; (2) the expectation that attempts to provide a true description of nature will result in one uniquely correct account is founded on an outdated theological-metaphysical thesis that the world was designed and created by a language-speaking agent, to whose privileged account the uniquely correct (true) account of science must correspond. One can be a realist about the world, I argue, without supposing that there is one uniquely correct description of it, just as one can be a realist about the past without believing that there is one uniquely correct account of historical events. This is not to say that expectations of convergence, consensus, or unification in science are always unfounded, only that they may not be necessary. Science seeks to solve different sorts of problems for which different tools may be required. In place of the metaphysical conception of objectivity as correspondence to the one privileged account of things, I agree with those who advocate for a social conception of objectivity as a value-driven methodological ideal for the attainment of reliable beliefs.

In substitute for the thesis that science aims for a uniquely correct (literally or objectively true) account of the world, I argue that a more pragmatic version of realism (like Giere's (2006) model-based perspectival realism) is a better fit for the fact that science relies so heavily on metaphorical language and thinking. Science is a creative activity in which humans use metaphors to construct models of the phenomena in question. Metaphors facilitate analogical reasoning from systems better understood to those still obscure. Our models and theories must be accurate pictures or representations of reality, must guide us to successful prediction, or allow us to successfully control and manipulate the world. They cannot do this unless they are in contact with some real systems or things outside of the models or theories themselves. In some instances, scientists may indeed seek to provide a literally true account of things, and if in doing so they rely on metaphor that cannot be replaced with literal language without losing the content of the theory or the insightful explanations that they make available, then this presents a problem for the realist. But in other cases it is obvious that scientists are really less interested in providing a literally true account and are just trying to build adequate models, they are trying to "puzzle solve," as Kuhn (2012) described the activities of "normal science"; or they are attempting to manipulate and intervene in some system, in which case it will matter less that they resort to metaphor to achieve their ends, so long as the metaphors help them to successfully get the practical results they are after. A number of contemporary projects in synthetic biology and biomedical research concerning cells especially are of this interventionist-engineering variety, activities which are often described as "technoscience."

If metaphors are a powerful and legitimate tool for scientists to employ, it cannot be denied that like other scientific instruments they can also introduce distortions and inaccuracies into our understanding. Associations carried along by metaphors from the original source domain can act as artifacts or misleading suggestions about the object under study. For instance, cells may communicate with one another by means of various signals, that are transduced and processed internally along signaling pathways and circuits, but as scientists have increasingly come to recognize, these individual pathways are not static entities like rigidly soldered electronic circuits at all, they are really processes, consisting of increased concentrations of proteins, peptides, and other molecules within the highly dynamic environment of the cell. Nor are these signaling pathways well insulated from the occurrence of "crosstalk" between them. In fact, researchers in the area now believe that all pathways are really "networks," and moreover, what was originally perceived as disruptive crosstalk is proving to be a powerful means by which cells coordinate and integrate signals. As so many writers on the question of science and metaphor like to say, apparently quoting the cyberneticists Norbert Wiener and Arturo Rosenblueth, "The price of metaphor is eternal vigilance." But the moral to draw from this is not that metaphor is to be avoided altogether in science, for it is too powerful a tool to discard; but rather that like any scientific instrument it requires careful and attentive use, recognition of its limitations and potential for creating artifacts, and improvement and calibration.

The issue of scientific metaphor tends to get written off as a soft kind of general philosophy of science question, of lesser importance than the hard questions of analytical philosophy of science regarding the logic of scientific method and reasoning for instance. This is unfortunate, given how pervasive metaphor is in science. Understanding what work a particular metaphor is doing in any particular area of scientific research is not trivial, it requires close attention to the science and the scientists in question, not just to how they talk, but to how they think, how they set up their experiments, conceptualize their observations, and conduct their research programs in general. Metaphors like the cell factory and cell signaling show that metaphors can be much more than expendable figures of speech used for communicating with the public; they form the constitutive heart of significant research programs in bioengineering, synthetic biology, biomedical research, and drug discovery. They act as guiding visions of what scientists hope to achieve–without the metaphors the expenditure of time, money, and effort on these projects would make little sense. While one is free to insist that scientists could *in principle* get by without the metaphors they employ, the fact is the metaphors are a central element of the scientific process, every bit as important as the material instruments and microscopes with which they investigate and create understanding of the world; and if we are to understand correctly how science operates, we must give due consideration to all the tools on which its many achievements rely.

1

The Early History of Cell Theory: The cell as empty chamber, building stone, and elementary organism

The changing nature of the cell concept and of concepts immediately related to it—from the container (the "cells" of Robert Hooke's cork, 1667) to the contained (the *Energide* of Sachs, 1892); from *Gallerte* (Treviranus, 1816) or *Schleim* (Schleiden, 1838), with or without *Körperchen* (Purkinje, 1836) or *Körnchen* (Valentin, 1835), to *Cytoplasma* (Kölliker, 1867) and *Kernplasma* (Strasburger, 1879); from "sarcode" (Dujardin, 1867) to universal *Protoplasma* (Cohn, 1850); from the "substance glutineuse, simple et homogène" of Dujardin to the immensely complex heterogeneous system which we know today—should serve as a permanent warning against a belief in the fixity of concepts, or in their value at any moment in time, save as a means of communication, or rapid reference to the present state of knowledge—Laurence Picken (1960, 1).

1. Introduction

This chapter provides a history of some of the early developments of the cell concept beginning in the seventeenth century and the cell theory in the nineteenth century up to the early twentieth century. This will be a highly selective history, focusing chiefly on the metaphorical language used to talk and think about those things we today so casually refer to as "cells." Because this will be a philosophical history, I beg the indulgence of professional historians of science, who will no doubt shudder at my lack of attention to important issues regarding material, technological, institutional, political, social, and other factors that are of course crucial to a complete understanding of the developments under discussion. My aim, however, is to illustrate that how to talk and to think about cells has always been just as important an issue as how to physically investigate them by means of material technologies and techniques. But before we begin our history of metaphors that have informed the cell concept and the cell theory, we must say a few words about what metaphors are and how they differ from similes and other related concepts.

2. Metaphors, similes, analogies, and models: a brief account

Metaphor and simile are figures of speech used to draw a comparison between two things. A *simile* typically employs the terms "like" or "as" and attempts to create a vivid image. For instance, "I slept like a log" or "She's as busy as a bee." In both cases it is understood that the intent is to assert that the two things in question are similar in some particular respect. The first suggests that while sleeping I was similar to a log in that I was silent and still. The second suggests that the person in question was as active as a bee flying rapidly and incessantly from one flower to the next. With a simile it is clear that the things being compared remain distinct and that there is no intent to assert an identity in all or even any essential features.

The *Oxford English Dictionary* defines a *metaphor* as "A figure of speech in which a word or phrase is applied to an object or action to which it is not literally applicable."[1] In contrast to similes, metaphors make comparisons without using the terms "like" or "as" that would indicate the two things in question share only limited resemblance. The statement "Life is a highway," for instance, invites us to compare the respects in which our life is similar to a highway. Both have beginnings and ends; both have unexpected twists and turns; we may encounter roadblocks, etc. But metaphors also encourage us to think of two things, not just as similar in some superficial respects, but as identical in some deeper fashion. So while we all understand that life is not really identical to a highway, when we use the metaphor we do tend to think of life as a *kind of journey*, and it is this underlying identity that makes the metaphor effective. Likewise, when scientists say things like "Genes are the units of hereditary *information* transmitted from one generation of organisms to the next," they are using the term "information" to describe the material molecules of DNA that get passed from one generation to the next. They are not simply saying DNA is *like or similar to* information, they mean DNA is in some essential sense a form of information. Information was traditionally associated with language, either spoken or written, and because the metaphor of information has also been associated with the metaphorical description of DNA as a genetic "code" (with triplets of nucleotide "letters" serving as "codons"), it has become quite natural for us to think of genetics and other aspects of biology as involving forms or types of information.[2]

The key difference, therefore, between a simile and a metaphor is that metaphors encourage us to think of two things not just as similar in some nonessential properties, but as identical in some important essential sense. Both are used in science to draw *analogies* and to facilitate *analogical reasoning*. This is the intellectual process whereby, on the basis of a perceived similarity between two objects or systems, we transfer our knowledge and understanding of one with which we are familiar, to another about which we are less familiar. Analogical reasoning is premised on the assumption that if two systems are similar in one or a few properties, they may also be similar in others yet to be discovered (Bartha 2013).

Philosophers of science have been very interested recently in the role played by models in scientific inquiry. *Models* are an important element in how scientists attempt to represent the world and understand how it functions. Models take many forms: physical, mathematical, pictorial-diagrammatic, and linguistic, and they also frequently (though not always) involve forms of analogical reasoning. Some analogical models, such as the billiard ball model of a gas or Sewall Wright's adaptive landscape model of geneotype fitness, can be expressed by either simile or metaphor. Many scientific models and theories have their origins in metaphor, e.g., Darwin's theory of natural *selection* or the electromagnetic *wave* model of light.[3] The key point for now is that metaphors are a powerful aid to reasoning by analogy. Detailed discussion about how scientific metaphors work will be taken up in chapter 5. I turn now to the history of metaphor in the creation of the cell concept and the cell theory.

3. Origins of the cell concept

The term "cell" was introduced into the natural sciences by the English naturalist and polymath Robert Hooke (1635–1703) in his account of observations of nonliving and living matter with a compound microscope, the *Micrographia* (Hooke 1665). The book contains sixty chapters, each devoted to observation of a specific material (the last three actually concerning Hooke's telescopic observations of the moon and stars). Chapter 18, titled "Of the schematisme or Texture of Cork, and of the Cells and Pores of some other such frothy Bodies," describes his observations of sections of dead cork plant. Hooke speaks of the porous nature of the material: "I could exceeding plainly perceive it to be all perforated and porous, much like a Honey-comb, but that the pores of it were not regular, yet it was not unlike Honey-comb in these particulars" (113). (See figure 1.1.) Although many popular accounts assert that Hooke was led to describe these small structural units as cells because they reminded him of the small rooms occupied by monks in a monastery, the actual text reveals that Hooke was in fact making a comparison to the polygonal cells of beeswax. The cell of honeycomb is itself a metaphor likely drawn from comparison to the small rooms of monks, so that Hooke's biological cells are a twice-borrowed metaphor. In describing the porous nature of cork tissue, Hooke uses alternately the terms "pores," "boxes," and "cells."

> Next, in that these pores, or cells, were not very deep but consisted of a great many little Boxes, separated out of one continued long pore, by certain Diaphragms . . .

A pore, as defined by the *Oxford English Dictionary*, is a "minute opening in surface, through which fluids may pass," and from the quotation above it seems that Hooke intended to use the terms pore and cell interchangeably, so that a cell in his original conception may in fact be composed of several smaller "Boxes." However, his usage of the term appears inconsistent, as the following examples illustrate:

> Our *Microscope* informs us that the substance of Cork is altogether fill'd with Air, and that the air is perfectly enclosed in little Boxes or Cells distinct from one another.
> . . . the whole mass consists of an infinite company of small Boxes or Bladders of Air . . . (113–14)

The pores are likened to "channels or pipes through which the *Succus nutritus*, or natural juices of Vegetables are convey'd, and seem to correspond to the veins, arteries and other Vessels in sensible creatures" (114). Hooke comments on the extremely minute size of these cells, pores, or boxes, suggesting that they may be too small even to allow the hypothetical atoms of the ancient Greek philosophers to pass through.

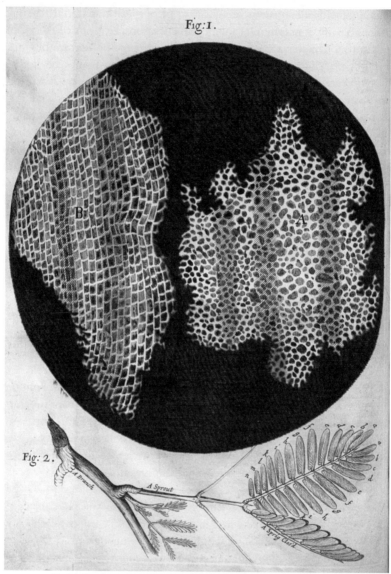

FIGURE 1.1 Hooke's first depiction of cells. "Observ. XVIII. Of the Schematisme or Texture of Cork, and of the Cells and Pores of some other such frothy Bodies." (From Hooke 1665, plate 11, fig. 1.)

The pith material contained within a feather quill is also described as being composed of "very small bubbles," each "Cavern, Bubble, or Cell" being "distinctly separate from any of the rest" (116).

In any case what does seem clear, whether he uses the term "cell," "box," "bubble," or "bladder," is that what grabbed Hooke's attention was a structural feature whereby the tissue was divided up into distinct spaces or units separated by a wall-like enclosure. Hooke's cell then was not our modern conception of the cell, for it was not intended to denote a living, physiological, or reproductive unit. Hooke was not attempting to articulate anything like a modern-day cell *theory*. He was not claiming to have discovered a universal principle of anatomy or physiology in plants, let alone in biology in general. He was merely describing a particular type of structure observable in some living material. Hooke's original notion of a cell, an empty space enclosed by solid walls, is an example of *catachresis*, the borrowing of an existing term to fill a gap in vocabulary (Soskice and Harré 1995, 303–4). He might have created a brand-new term for the structures in question. For instance, he might have called them "jexes," for the simple purpose of having some label by which to refer to them, and had he done so we can only speculate how things might have turned out differently.

Other philosophers or naturalists would take up Hooke's term and apply it to similar structures in other plants. For instance, in the late seventeenth century and throughout the eighteenth, a multitude of investigators armed with microscopes would describe similar structures in a variety of plant tissues. These spaces were variously called "cells," "bubbles," "bladders," "cavities," and "vesicles" by observers such as Marcello Malpighi (1628–94), Nehemiah Grew (1641–1712), Albrecht von Haller (1708–77), and Christian Wolff (1679–1754). Despite an agreement that these structures were to be seen in plant tissue, there was fundamental disagreement about whether these cells were positive entities in themselves or merely empty spaces or voids in an otherwise continuous material.[4] In other words, it wasn't clear whether these cellular spaces should be considered as foreground or background. Was the cell a *thing* or the mere *absence* of things? Some compared the presence of these cells to the bubbles in a foam, the froth of beer, or the holes in lace, while others regarded them as real and distinct entities. Use of the term "cellular tissue" or "*Zellgewebe*" by eighteenth- and early nineteenth-century writers in relation to animal anatomy further confuses the issue, for this term was used not with the modern-day notion of cells as distinct units in mind, but to describe a web-like appearance in connective (areolar) tissue formed by a network of fibers (Wilson 1944; Baker 1948, 112–14).

4. Origins of a cell theory

The microscope was an ingenious bit of technology that provided unforeseen powers of observation for the curious minded, but it was insufficient for the creation of what we know as the cell theory. Even astute observers like Antoni van Leeuwenhoek (1632–1723), the discoverer of the minute world of the *infusoria* (ciliates and bacteria), refrained from proposing that his observations had uncovered a common unit or principle of structure underlying all living forms.[5] Nor was the microscope entirely necessary for this purpose. For at this time many writers were beginning to speculate their way toward philosophical theories of the nature of plants and animals and life in general. Georges Louis LeClerc de Buffon (1707–88) for instance suggested that plants and animals are composed of "little organized beings," which are themselves composed of primitive and incorruptible living atoms (Buffon 1749, 24), and the German *Naturphilosoph* Lorenz Oken (1779–1851) speculated that all animal flesh is composed of smaller *Urthiere* or infusoria (Oken 1805, 22). Many commentators have made the case that the theory of a universal principle of plant or animal structure was as much the result of prior philosophical ideas—chief among them atomism or corpuscularianism—as it was improvements in microscope technology.[6] And yet it cannot be denied that improvements in the optical design of microscopes between the years 1830–40 to correct for spherical and chromatic aberrations significantly assisted efforts to identify an underlying and unifying principle of anatomy and physiology. Prior to these improvements, observations of minute elements in living tissue were disputed as artifacts confounded by imperfections in the lenses, halos of light being mistaken for "globules," for instance.

This early period of microscopical investigation also saw a diversity of alternate terms in circulation. While Stefano Gallini (1756–1836) used the term "cell" to denote a precise anatomical unit in 1792, others continued to employ the terms "bubble," "vesicle," "bladder," and "globule" (Dröscher 2014a). And in addition to these, the histologist Jan Purkyně (1787–1869) used the German terms "*Körnchen*" (little kernel or seed) and "*Kügelchen*" (little sphere) (Harris 1999, 86). So what advantage did the term "cell" have that would explain its eventual rise to dominance? I would argue that it was largely accidental. The cell concept does have certain merits, but as became clearer as microscopical investigation of animal anatomy and unicellular protozoa in particular advanced throughout the nineteenth century, the cell metaphor, with its suggestion of a clearly defined and rigid wall, is both inadequate and misleading as a universally applicable term. What the cell concept did provide was a useful search image for early synthesis-minded scientists who were looking to discover some unifying principle of design by which to arrange the plant and animal kingdoms. By emphasizing a unit surrounded by a distinct wall or boundary, the cell concept helped to focus the attention of investigators while they observed various specimens under the microscope. Of course the same can be said of all the other terms mentioned, insofar as they all highlight a discrete unit or entity. Harris (1999, 86f) mentions that whereas *cellula*, from which we get "cell," is Latin for an empty interior into which things can be put, the German equivalents for *korn* or *kernel* favored by people like Purkyně and his students, denotes a solid body from which living organisms develop.

For whatever reasons, "cell"—or rather the German equivalent *"Zelle"*—was the term used by the botanist Matthias Schleiden (1804–81) and his zoologist colleague Theodor Schwann (1810–82), and it is to them that the first articulation of the cell theory is most frequently credited.[7] Schleiden, like many others at this time, was interested in identifying a fundamental unit of plant biology, a search that involved arriving at a proper conception of the plant individual.[8] For Schleiden, who believed like many others that plants are compound organisms aggregated from simpler units, this was the individual cell (Elwick 2007). In many types of plants, well-defined cell walls are reasonably easy to identify with a microscope, especially if one is looking for them (a charge made by later critics of the cell theory). Schleiden also capitalized on the visible presence of the nucleus–which had been described by Robert Brown (1773–1858) in 1833—to help identify individual cells in the composition of plant tissue. But Schleiden was not primarily looking for a structural or anatomical principle common to all plants; he was looking for a physiological and developmental unit or individual (note that the title of his monograph is *Contributions to Phyto-genesis*). As he wrote in 1838,

> The idea of an individual, in the sense in which it occurs in animal nature, cannot in any way be applied to the vegetable world. It is only in the very lowest orders of plants, in some *Algae* and Fungi for instance, which consist only of a single cell, that we can speak of an individual in this sense. But every plant developed in any higher degree, is an aggregate of fully individualized, independent, separate beings, the cells themselves. Schleiden (1847, 231–32)

Schleiden declared next that "each cell leads a double life: an independent one, pertaining to its own development alone; and another incidental, in so far as it has become an integral part of a plant" (232).

So his thesis that all plants are composed of cells was more radical than the simple claim that plants are based on a cellular construction. For this much could be granted while still insisting (as some later botanists did) that the plant as a whole is the proper individual, which just happens to divide itself up into cells as it grows and develops. Such was the opinion of later botanists like Anton de Bary (1831–88), who is famous for having said, "It is not the cells that make the plant, but rather the plant that makes the cells."[9] As the passage from Schleiden makes clear, right from the early beginnings of the cell theory, cells were conceived as more than just inert structural units (building stones or *Bausteine*, as they would frequently be called in the twentieth century). In fact, Schleiden asked in the same essay, "How does this peculiar little organism, the cell, originate?" (Schleiden 1847, 232). This consideration of cells as organisms in their own right will have significant consequences, as we shall see.

Schleiden believed that new cells arise within preexisting cells as an accretion of material around the cell nucleus (or *cytoblast*, to use his term). But his insistence on the priority of the cell as the ultimate living individual in plants would have great and lasting influence through its effect on his colleague Theodor Schwann. Both men were studying in Berlin at the time with the renowned animal physiologist Johannes Müller (1801–58). Schleiden told Schwann of his idea that the cell is the fundamental individual from which all plants are composed, leading Schwann to pursue the possibility that cells are the true organic individuals in the animal kingdom as well. Schwann had earlier noted a similar cell-like construction in some animal tissues, and Plate 1 of his work on the cell theory (Schwann 1847 [1839]) compares the cells of onion tissue to the cellular structure of the developing notochord and cartilage from fish, toad, frog and pig (see fig. 1.2). But the great diversity in appearance of mature animal microanatomy impressed upon him that many mature animal tissues do not appear to be composed of clearly defined cells. The chief significance of Schleiden's cell theory of plants for Schwann was therefore the suggestion that the cell is a uniform principle of animal tissue development (*Entwicklung*), so that even those adult animal tissues that do not appear to be composed of isolated, independent cells can be recognized as the result of cell fusion and subsequent modification. It was the cell as the universal unit of development, not necessarily of anatomy, that allowed Schwann to unite the plant and animal kingdoms.

Schwann presented his case for the primacy of the cell by first documenting examples of anatomical units easily identifiable as isolated,

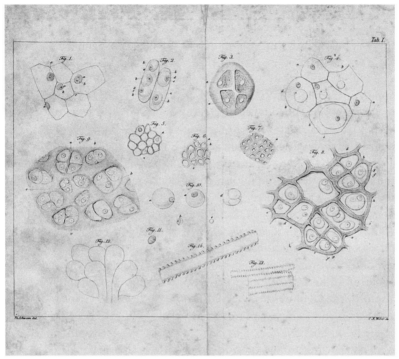

FIGURE 1.2 Schwann's comparison of plant and animal cells. Figs. 1–3, and 14 are of plant cells, the rest are animal fetal and larval cells. (From Schwann 1839.)

independent cells (e.g., lymph, blood, mucus, and pus cells), moving next to "independent" cells united into continuous tissues (epithelia etc.), to coalesced tissue cells (cartilage, bone, teeth), and lastly "tissues generated from cells, the walls and cavities of which coalesce together" (muscle, nerves, and capillary vessels). Maintaining that the cell is the universal principle of plant and animal development meant more than simply identifying it as a general anatomical structure; it entailed giving it special priority as the most fundamental living unit or agent. As Schwann wrote in 1839, "Each cell is, within certain limits an Individual, an independent Whole" (Schwann 1847[1839], 2). The qualification "within certain limits," on the other hand, was necessary to note the difference in the degree of integration and interdependence of the cells of plants and animals, respectively. Schwann wrote,

> This resemblance of the elementary parts has, in the instance
> of plants, already led to the conjecture that the cells are really
> the organisms, and that the whole plant is an aggregate of these

organisms arranged according to certain laws. But since the elementary parts of animals bear exactly similar relations, the individuality of an entire animal would thus be lost (190).

For this reason, Schwann allowed that in the case of animals the whole
appears to be more than the sum of its parts. Although a higher animal
is composed of individual cells, it is not simply an aggregate of cells, as
a sand pile is an aggregate of grains of sand, he acknowledged (2), and
unlike many plants, a complete animal cannot typically be grown from
a small cutting of the whole.[10] But Schwann was declaring that animals
too are constructed from and by these elementary organisms, the cells,
the ultimate agents of life.

According to Baker (1948, 103) it was Schwann who first used the
phrase "cell theory" (*Zellenlehre*), by which he meant the thesis that the
process of cell formation was a general principle of construction (*Bil-
dungsprinzip*) for all organic products, vegetable and animal alike. In
addition to this largely empirical or inductive generalization, Schwann
also proposed a more speculative "Theory of the cells" (Schwann 1847,
186–215), with which he hoped to explain how cells develop and reproduce. In slight contrast to Schleiden, Schwann thought that cells
were capable of forming outside of preexisting cells, around free nuclei,
which he believed emerged in a nutritive fluid in a process akin to the
growth of crystals.

Studies on the cellular basis of animal embryogenesis by Martin
Barry (1802–55), John Goodsir (1814–67), and Robert Remak (1815–
65) helped to displace Schleiden and Schwann's account of how cells
reproduce. In a series of observations (published 1851–55) on the development of the frog, Remak demonstrated that the fertilized egg and
early embryo cells all replicate by the process of cleavage, one cell dividing in half to produce two (Remak 1855). The pathologist Rudolf
Virchow (1821–1902), without expression of proper credit to Remak,
popularized this with the slogan "Omnis cellula e cellula" ("Every cell
originates from a previously existing cell"). This development is often
considered to be the final culmination of the classical cell theory, ex-
pressible in three main theses:

1. All organisms consist of one or more cells.
2. The cell is the ultimate unit of life, development, and
   reproduction.
3. All cells come from preexisting cells by division.

We will, however, shortly consider a fourth thesis commonly attributed to the cell theory, the more contentious claim that:

4. The life of an organism is nothing more than the sum of the properties and activities of its component cells.

It is important to note that although the cell metaphor emphasizes a structural feature of living material (CELLS ARE SPACES DEFINED BY SOLID WALLS), as Schleiden and Schwann employed it, the cell denoted an elementary living particle with a very plastic ability to transform itself into a diverse range of morphological and physiological results. The cell metaphor was actually ill suited for these developmental-physiological purposes, but it did provide a useful search image that allowed early observers to notice an accordance of structure across plants and animals. But for the purpose of accounting for a truly universal *physiological* and *developmental* principle, Schleiden and Schwann both conceived of the cell as an elementary organism, thereby introducing the metaphor that CELLS ARE ELEMENTARY ORGANISMS. Why do I call this a metaphor? A metaphor recall is a figure of speech whereby one subject is described in terms more naturally or conventionally fitting of another. When Schleiden and Schwann referred to the cells of plants and animals as organisms, they were saying something quite novel and striking for the time, for it was not then common usage of the term "organism" to be applied to parts of what were standard examples of organisms, e.g., a human, a dog, or a plant (Cheung 2010). This innovative use of language would have significant implications, as we shall see.

Parnes (2003) argues persuasively that the cell represented for Schwann a natural (i.e., nonvitalist) agent capable of carrying out the physiological functions of the various living tissues and organs. Nutrition, respiration, etc., Schwann insisted, are processes carried out not by the organism as a whole, but by the individual cells. And what is sufficiently accounted for at one level need not be presumed to recur at another, rendering appeals to a harmonious vital principle functioning at the level of the organism as a whole unnecessary.

At this stage the cell theory consisted of a structural-anatomical thesis and a developmental-physiological thesis. It would eventually also include an evolutionary-phylogenetic thesis (developed principally by Haeckel in the latter half of the nineteenth century) and a genetic thesis of heredity (developed toward the end of the nineteenth and beginning of the twentieth century).[11] The anatomical thesis was vividly expressed by the botanist Franz Unger (1800–70), whose popular *Botanische Briefe* (1852) described the cell as the building stone (*Baustein*) from which plants are constructed (18, 20, 21, 25). Unger spoke of the repeated stacking of cells in some plant tissues as forming a *Mauerwerk*, a wall of stones or bricks (12, 14, 19, 26; see fig. 1.3) This is an appearance characteristic of epithelia in both plants and animals. In fact, one type of epithelial cell (simple squamous) is commonly known as "pavement epithelia." Unger did, however, note the inadequacy of this analogy, as it attributes to them more independence than he believed them to have (19, 24–25).

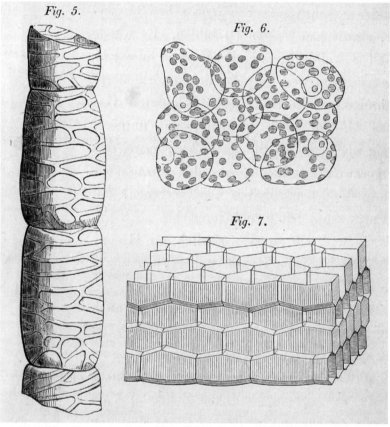

FIGURE 1.3 Plant tissue composed of multiple cells (a "wall composed of multiple building stones"). (From Unger 1852.)

5. Cell theory as social theory

Though a conception of cells as the modular building stones of plant and animal anatomy possessed a natural plausibility, cells continued, nevertheless, also to be thought of and characterized as organisms. In fact, Unger, in addition to speaking of the plant cell as a building stone also described it as an ever-busy *Spagiriker* or alchemist (38), thus exchanging (or complementing) an artifact metaphor with a more active agent metaphor. A building stone, a brick, and a cell (in the literal sense) are all artifacts created by living organisms (the first two exclusively by humans, the last by humans and bees), and organisms are agents, which means they can act and effect change on inanimate matter and other agents. When agents act on one another they become social beings, and from this social interaction new properties and modes of behavior emerge that are not possible when agents act alone in isolation from one another. This is an important theme of cell theory obscured by characterizations couched in the language of artifacts such as cells and building stones.

As Jan Sapp (1994, 36) notes, throughout the nineteenth century (especially the latter half) the cell theory took on the aspect of a distinctively social theory. Both Schleiden and Schwann, we saw, regarded cells as minute organisms. So it is hardly surprising that some people began to think of plants and animals as analogous to human societies, cities, or states, comprised of these tiny cell-individuals.[12] The English philosopher Herbert Spencer (1820–1903) was one of the first to do so, when in 1851 he compared the cells of the animal body to the "infusorial monads" or protozoa –which Leeuwenhoek had been first to see. Noting a division of physiological labor in the arrangement of cells of the body into various tissues and organs, he concluded that "we are warranted in considering the body as a commonwealth of monads" (Spencer 1868, 493; see Elwick 2003, 2007, 2013). Spencer made wide use of this analogy between the cells of a living body and the citizens of a modern society—an idea he referred to as the "Social Organism"— in his broadly popular and influential writings on social and political philosophy.

At around the same time, Rudolf Virchow began describing the human body as a "Cell-State" (*ZellenStaat*). The living organism, Virchow wrote, is "a society of cells, a tiny well-ordered state, with all of the accessories–high officials and underlings, servants and masters, the great and the small" (Virchow 1958 [1858a], 130). Such comparisons were possible not only because one conceived of cells as elementary organisms, but because the physiological labor of the body can be thought of as being divided up into distinct tissues and organs where specialized cells perform distinct and specific tasks.[13] The cell-state metaphor served Virchow in two separate ways: one scientific, the other political. The political function was to emphasize the rights of the individual in the context of mid-nineteenth-century efforts to establish a unified Germany from the several distinct German states and principalities. Some members of the traditional land-owning and noble classes were pushing for a centralized and conservative arrangement that best served their own interests, and these tended to invoke the metaphor of the social organism wherein each member of the body has its proper place and function (Mazzolini 1988, 81–95). According to this image, any efforts to rearrange the natural order were characterized as a sickness that could only jeopardize the well-being of the whole. Virchow was a cofounder of the progressive *Deutsche Fortschrittspartei* and served in the Prussian Diet and as a representative in the Reichstag. In response to the rhetorical language that painted the state as an inegalitarian hierarchy of privileged classes ruling over subservient parts, Virchow used his status as a respected physician and research scientist to turn these popular metaphors on their head. Invoking the cutting-edge science of cell theory, Virchow informed his audiences that the animal body is in fact a "free state of individuals with equal rights though not with equal endowments," it is a "Federation of cells," a "democratic cell state."[14] His choice of metaphors clearly reflected his republican and liberal political sentiments.

The scientific function of the "cell-state" metaphor, on the other hand, was to highlight the primary importance of the cell as the fundamental unit of life. Because the idea that society is like a large organism composed of individual people was already familiar, he could use the reverse analogy–that the organism is itself a kind of society—to emphasize that just as society is composed of individual people, so the animal body is constituted from individual cells. This allowed Virchow, whose chief interest was in establishing pathology on the basis of cell theory, to make the case that just as an epidemic of disease spreading through a society is a matter of sick individuals, so illness in the organism is a matter of dysfunction of individual cells, not of privileged classes alone such as the blood or nervous system as the then-dominant humoral theory of disease supposed. Virchow's influential lectures on *Cellular Pathology* explained that every individual organism of a significant size "represents a kind of social arrangement of parts" (1858b, 12–13). It is necessary, however, Virchow remarked, to consider the influences neigh-boring cells can have on one another, an aspect he marked by speaking of "cell territories." So although cells may be the fundamental units of life, understanding the nature of any specific cell required considering it within its larger social context, as a member of a particular form of tissue or organ.

This theory of the cell-state was taught at universities throughout the German territories, but the young medical student Ernst Haeckel (1834–1919) learned it directly from Virchow at the University of Wurzburg in the 1850s (Weindling 1981, 99–155). Shortly after com-pleting his medical training in 1857 (with a dissertation on the histology of a species of river crab) Haeckel read the first German translation of Darwin's *On the Origin of Species*.[15] Haeckel was a pioneer in the com-bination of evolution and cell theory and was particularly interested in finding evidence of the origin of multicellular organisms from unicel-lular ones (those he called protists). Following Virchow's lead, Haeckel frequently referred to the plant and animal organism as a "state of cells" (*ein Staat von Zellen*), a "cell-society or cell-state" (*Zellen-Gesellschaft oder Zellen-Staat*) (Haeckel 1866, I, 264, 270), and even "a republican cell-state" (*ein republikanische Zellenstaat*) (Haeckel 1868, I, 246).[16] But whereas Virchow had used the cell-state metaphor as an anatomi-cal and physiological thesis for thinking about the relationship between the organism as a whole and its cell-parts, Haeckel as a confirmed evo-lutionist used it to speculate on how modern-day complex cell-states represented by animal bodies might have evolved from more primitive sorts of cell societies.

In addition to being one of the most influential proponents of evo-lutionary thought in the nineteenth century, Haeckel was also a leading investigator of microscopic life, with many respected volumes devoted to the single-celled organisms, which he collectively called the Protista (e.g., radiolaria, amoebae) and to marine invertebrates (e.g., sponges, corals, and jellyfish). He claimed to have observed single-cell organisms so simple in structure as to lack a nucleus, a structure previously con-sidered characteristic and in fact diagnostic of the cellular form of or-ganization. Haeckel called these simple anucleated units "cytodes"—to distinguish them from true cells—and introduced the taxonomic class name *Monera* as label for these presumably most primitive and ancient organisms.[17] He proposed that isolated "hermit cells" (*Einsiedlerzellen*) of the Moner variety gradually gave up their solitary ways to form sim-ple colonies similar to present-day protist colonies like Synura and the Volvocales.[18] These simple colonies of cells he compared to monastic communities in which there is a simple social organization and limited division of labor.[19] Increased division of labor would result in more specialized and differentiated cells, so that eventually as the interdependence of cells reached the stage of higher organisms none could survive on its own, each having become reliant on the specialized functions of its neighbors.

By 1866 Haeckel had already proposed a hierarchical theory of biological individuality ranging from the individual cell, the organ (composed of cells), up to individual (multicellular) organisms, some of which, like the colonial Siphonophores, exhibit a curious mix of individuality and community.[20] Haeckel also made a distinction between those cell-states in which the individual cells enjoy a greater degree of freedom and autonomy from those wherein their activity is more restricted and regulated by a centralized system of control. The former, best represented by plants, he referred to as "cell-republics," while the latter, which he called "cell-monarchies" are represented by animals. The vertebrates, because they possess a centralized nervous system that regulates and integrates the activities of all the cells, tissues, and organs, represented for Haeckel the most advanced cell-states, and these he likened to his own late nineteenth-century German *Cultur-Staat*.[21]

A major implication of the cell-state theory was that cells are essentially autonomous individuals, despite their often being intimately conjoined—morphologically and physiologically speaking—with their neighbors, as is the case with epithelial cells. This belief in the fundamental autonomy of cells had significant effects on Haeckel's thinking, particularly as an embryologist. For instance, in 1866 he carried out investigations into the development of several species of Siphonophora, an order of marine invertebrate of the class Hydrozoa, which although resembling jellyfish (Scyphozoa) are actually colonies of medusoid and polypoid individuals. These colonies display such a remarkable degree of division of labor and specialization that it was a matter of dispute whether they consisted of one individual organism with several specialized parts or a colony of several highly differentiated individuals.[22] Haeckel and some other German biologists actually referred to the Siphonophora as *"Staatsquallen"* (State-jellyfish), a term that emphasizes their colonial nature. They represented, in Haeckel's hierarchical theory of relative individuality, what he called a *Stock* or *Corm*, an individual of the sixth order. As he observed under the microscope the division of the single fertilized egg cell into two, then four, then eight cells, and so on, to form the developing embryo, he was struck by the creeping movements of the individual cells as they organized themselves into what would later become the more specialized tissues and structures of the mature medusoid and polypoid individuals. "One could be struck by the thought" he wrote, "that the whole body of the two-day-old Siphonophore larva, a mere spherical aggregate of large, hyaline, amoeboid cells, could be compared to a colony of amoebae" (Haeckel 1869, 73; see fig. 1.4). And as he watched the amoeboid movements of these embryonal cells (blastomeres), it occurred to him to try the experiment of separating some of them from the embryonal heap to see whether they would develop into a complete individual on their own.[23] That several of these artificially cleaved blastomeres not only survived, but developed into not-quite-complete but rudimentary Siphonophore stock confirmed his belief that the cells of highly organized animals are very likely the descendants of formerly isolated and autonomous cells from a more ancient time.

With Haeckel we see that cells were not only given priority as the fundamental anatomical and physiological unit, as they had with Virchow and other early cell theorists, but they are also considered to be an important evolutionary unit. This allowed Haeckel to use the history of the development of human societies as an analogical guide to understand the evolution of multicellularity and subsequent episodes known today as the "major transitions in evolution." As he believed: "The cultural history of humanity explains to us the organizational history of the multicellular organisms" (Haeckel 1879, 37). The analogy provided by the "theory of the cell-state" was so central to his approach that he claimed the whole understanding of biology depends upon "the political founding thought of the cell theory" (*Dieser politische Grundgedanke der Zellen-Theorie*), viz. that cells are semiautonomous citizens working together through a division of labor and centralization of activity (37). In fact, Haeckel took it as more than just metaphor: "This perfectly true and often employed political comparison is no remote symbol, but rather demands real validity; the cells are real [*wirkliche*] citizens" (36). This idea wove together for him the separate facts of morphology, anatomy, embryology, evolution, and the social history of humankind. The law of the division of labor (*Arbeitstheilung*) was for Haeckel a general law of progressive development or evolution that could explain the development of modern complex society (a *Culturstaat*) and the evolution of tightly integrated multicellular organisms as homologous developments. His oft-maligned endorsement of the biogenetic law, that an organism's ontogeny is a brief and shortened recapitulation of the evolution of the branch of life to which it belongs, was a further expression of this idea. We will discuss again at greater length this particular pair of

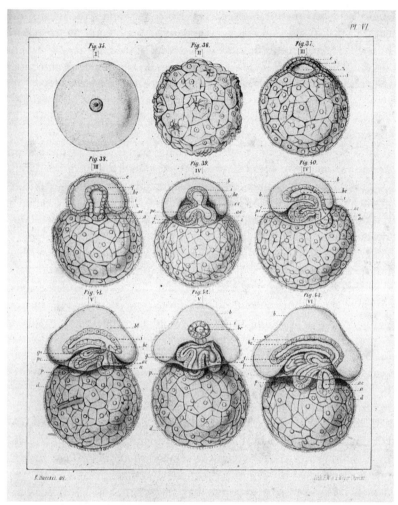

FIGURE 1.4 Blastomere cells from 1–6-day-old siphonophore embryo, from Haeckel 1869. Fig. 36 showing the amoeboid cells. (From the collections of the Ernst Mayr Library, Museum of Comparative Zoology, Harvard University.)

metaphors that THE BODY IS A SOCIETY OF CELLS & CELLS ARE INDIVIDUAL CITIZENS.

6. Protoplasm and the crisis of the cell concept

The cell, as a term and a metaphor, draws attention to the boundary encapsulating a fragment of living matter. It brings focus on the prison cell rather than the cell's content, the occupant or prisoner living inside the cell, which was typically described as a semifluid, mucus-like slimy substance with the capacity for contractility or "irritability." Initial attention to the cell wall was natural enough, as many plant cells possess a rigid and comparatively thick cellulose wall. But as many nineteenth-century microscopists were aware, there are plenty of examples from among the cells of animals and infusoria, which seemed little more than a gelatinous clump lacking any noticeable membrane, let alone a solid outer wall. Felix Dujardin (1802–60) dubbed the slimy material of which the infusoria or protozoa are composed "sarcode" or "living jelly" in 1835. Jan Purkyně (1787–1869) introduced the term "protoplasm" in 1840 for the ground substance of young animal embryo cells, a term that had also been applied in 1846 by the botanist Hugo von Mohl (1805–72) to the inner "slime" of plant cells. By the 1850s several investigators including Franz Unger and Robert Remak were suggesting that animal protoplasm and plant sarcode are one and the same substance. It was also in midcentury that the argument for designating the protozoa as organisms consisting of but one cell was made by Carl von Siebold (1804–85), Albert von Kölliker (1817–1905), and Haeckel, among others.

On the basis of their investigations of amoeboid creatures from the Rhizopoda and Myxomycetes (slime molds), respectively, Franz von Leydig (1821–1908) in 1857 and Anton de Bary (1831–88) in 1859 both concluded that a membrane is an inessential feature of cells, for it seemed unlikely that these ever-changing amoebae could manage to ooze about by extension of pseudopodia from their cell surface if they were restricted by an outer lining. Moreover, de Bary documented how, in Mycetozoa, free-moving reproductive "swarm cells" emerge from reproductive spores appearing to be little more than naked specks of protoplasm. In 1861 Max Schultze (1825–74) proposed a new definition of the cell based on the work of Leydig and de Bary and his own study of amoebae and animal muscle, stating that the cell is "a naked speck of protoplasm with a nucleus" (Schultze 1861; Reynolds 2008a). In this same year Ernst Wilhelm von Brücke (1819–92) published "Die Elementarorganismen," solidifying the argument for regarding cells as elementary organisms in their own right. But in calling them "elementary" organisms, it was not Brücke's intent to suggest that cells are simple or homogenous in structure, for he recognized that in order for the protoplasmic material to carry off all the vital functions it did— growth, metabolism, contractility/irritability, movement, and reproduction—it must possess a complicated internal organization of its own.

Rather than suggesting that a cell is elementary in the sense of being simple, he meant the cell was for the biologist what an element was for the chemist, a least complicated unit of life that could not be further reduced to any simpler organization while preserving all the characteristic properties of living matter. Although he saw no positive evidence in its favor, Brücke remained open to the possibility that future investigation might show the cell itself to be composed of yet smaller living units or even more elementary organisms.

Once released from its prison cell, protoplasm quickly began playing a more vital role in scientific discussions about the nature of life. This is illustrated in the fourth edition of William Benjamin Carpenter's (1813–85) influential *Manual of Physiology, Including Physiological Anatomy for the Use of the Medical Student*, which appeared in 1865. In a new preface introducing the reader to the protoplasmic theory of the cell, Carpenter explained why the new edition was required:

> It now appears to be conclusively established that the Cell, with its membranous wall, nucleus, and contents, is no longer to be taken as the primitive type of organization; but that the nearest approach to this type is to be found in the segment of 'protoplasmic substance' or 'sarcode' which forms the entire body of the lowest Animals:—and further, that the portion of the fabric of even the highest Animals which is most actively concerned in Nutrition, is a protoplasmic substance diffused through every part, its segments being sometimes isolated by the formation of 'cell-walls' around them. Hence the study of the life-history of the *Rhizopoda*, which their ordinarily minute size and transparence renders comparatively easy, comes to throw a most unexpected light upon the phenomena which occur in the innermost *penetralia* of the complex organization of Man (Carpenter 1865, vii).

T. H. Huxley (1825–95)—originally critical of the cell theory—became a great champion of the protoplasm theory. Huxley's influential essay of 1868 declared protoplasm "the physical basis of life." Whereas the original cell theory sought to unify all the various forms of life through a common morphological type and developmental principle, the protoplasm theory attempted to achieve this through the identification of a common substance or material. Protoplasm actually offered a threefold means of unifying all the different forms of animal, plant, and protozoan life: a unity of substance (the colorless, albuminoid, proteinaceous protoplasmic matter itself); a unity of function (irritability, nutrition, division, movement); and a unity of form (essentially a variably shaped speck of life-slime or jelly with the capacity to produce nonliving formed materials such as a cell wall or membrane, flagella, cilia, or hard encasing material like shell, bone, or nails). This was a physiological conception in keeping with the new understanding that the structural feature explicitly highlighted by the term "cell" was not an essential feature of life, as shown by many of the simplest organisms known to science.

Calls for the abandonment of the term "cell" soon followed, but despite several alternate candidates being proposed (e.g., corpuscle, bioplast, plastid), many scientists conceded that the term had been around so long and was so strongly associated with the important progress in anatomical and physiological science of the nineteenth century that any attempt to dispose of it would be futile and counterproductive.[24] It was also unnecessary, as it turned out, in any case: for the ascendance of the protoplasm theory did not replace the cell theory at all, as it initially looked as though it might. By means of a convenient chain of associations, the idea of protoplasm as an essentially shapeless living-jelly substance and the new conception of the cell as a "clump of protoplasm with a nucleus" united in an exemplar of the new protoplasmic theory of the cell.[25] This was the amoeba. The amoeba—more correctly the amoebae, for there is a diversity of cells capable of displaying the shapeless crawling form known as amoeboid—quickly became exemplary of the protoplasm theory and provided a bridge between the traditional cell concept (in terms of a morphological type) and the newer protoplasmic concept (in terms of an essentially amorphous substance). This revised protoplasmic cell concept, exemplified by the supposedly lowly amoeba, also worked well in the context of evolutionary theories about the community of descent and the rise of complex, orderly creatures from simpler, more primitive ones (see fig. 1.5). And because microscopes at that time were unable to reveal any obvious structure within the protoplasmic substance itself, it was commonly assumed to be rather homogenous. From this, some (e.g., Huxley) drew the mechanistic conclusion that life is ultimately reducible to the physical and chemical properties of the protoplasmic molecules, thus rendering unnecessary appeal to a vitalistic life-force.[26]

In this way the cell theory was able to survive the demotion in status of the specific feature from which it had derived its name. That it was able to do so was likely at least in part because the two researchers who were so intimately associated with it (Schleiden and Schwann) had spoken of the cell as a little organism, which allowed the cell concept

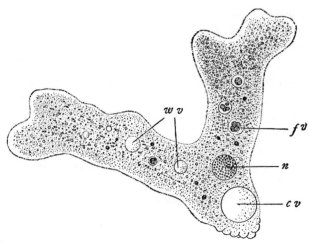

Fig. 3.—*Amœba Proteus*, an animal consisting of a single naked cell, × 280. (From Sedgwick and Wilson's Biology.)
    n. The nucleus; *w. v*. Water-vacuoles; *c. v*. Contractile vacuole; *f. v*. Food-vacuole.

FIGURE 1.5 Amoeba as exemplary cell. (From Wilson 1900.)

to be easily accommodated into the new era of protoplasmic thought. "Cell" is even less descriptive or fitting today, for who now even thinks of its original connotation when hearing or using the word? The term is today a dead metaphor because very few people, I suspect, ever think of tissue or simple organisms as being constructed of empty chambers defined by solid walls. The term "cell" has simply become a marker for whatever is meant by the expression "the fundamental unit of life." In a similar way, what physicists originally called the atom turned out not to be *atomos* (indivisible), but that hasn't stopped scientists from continuing to use the term. So while initially the metaphor may have determined the concept (i.e., the idea of a space defined by a rigid wall defined the cognitive meaning associated with the term), eventually the concept (fundamental unit of life) has come to dominate the term, which today is seldom even recognized to be a metaphor. But as we will see yet in ample abundance, after arriving at the claim "All living things are made of cells," biologists still need to describe what it is cells do and how they manage to do it, and this leaves plenty of room for new metaphors to thrive and compete for attention.

7. Parts and wholes: the cell vs. the organismal perspective

It has been suggested that the cell theory, being essentially an atomistic doctrine, is the inevitable result of an analytic approach in the life sciences (Woodger 1929, 209; see also Canguilhem 2008; Nicholson 2010). Whether cells are conceived structurally as "building blocks" (*Bausteine*) or physiologically as elementary organisms, the implication is that they represent modular units that enjoy a significant degree of autonomy and independence from one another and from the higher-level organisms of which they may be a part. Once this assumption is granted it would seem to follow that the properties and capacities of anything composed from these units (e.g., a plant or an animal) will be a simple additive or aggregative sum of the properties and capacities of its units. This is often expressed as a fourth thesis of the cell theory:

> 4. The life of the organism is the sum of the activities of its component cells.

The organism, it seems, is thereby reduced to its constituent cells and their activities and properties, leaving no vital property that is not a property of a cell or cells. Not surprisingly, this thesis has been the target of a great deal of criticism. In fact, expressed in such bold form it may have been more the creation of the critics of the cell theory than any of its advocates.[27] For instance, at the turn of the twentieth century the anatomist Martin Heidenhain (1864–1949) noted that it came in both an anatomical and a physiological form, which he called the "building stone theory" (*Bausteintheorie*) and the "theory of the cell-state" (*Theorie vom Zellenstaat*), respectively (Heidenhain 1907, 26, 54).[28] Haeckel, we saw earlier, was a strong advocate of the metaphor of the cell-state, and he also employed the *Baustein* metaphor to describe cells and plastids in books such as *Generelle Morphologie* (1866) and his extremely popular and widely influential *Natürliche Schöpfungsgeschichte* (1868), based on his lectures at the University of Jena.[29] The "Cell Theory" or "the Cell Doctrine," as it was commonly called, elicited two chief responses from its critics: (1) The cell is a mere part of the organismal whole, like an organ–it is not an autonomous whole in its own right; (2) The cell is not the most fundamental unit of life: some particle or corpuscle even smaller and more fundamental exists within the cell. In addition to these a third response is also possible: (3) The cell perhaps may be an elementary organism/individual, but it can be properly understood only in the context of the organism as a whole, so that consequently, in the case of plant and animal cells at least, cells do not exist in isolation and independent of the rest of the organism, and they are not in that case autonomous living units. Because the metaphor of the cell implies privacy, solitude, disconnectedness, this set up in the mind of many thinkers the problem of how to account for the harmoniously coordinated and functional nature of organisms as integrated wholes.

Its critics charged the cell theory with failing to provide an adequate description of anatomical, physiological, developmental, and taxonomic reality. Various sorts of organisms, tissues, and developmental phenomena have been appealed to as counter-examples to the claim that living bodies are composed of distinct cell units and that the vital properties of the organism are reducible to the vital properties of its component cells.

T. H. Huxley offered one of the earliest and most widely cited critiques of the cell theory in 1853, in a review of several works in anatomy and embryology. Huxley was chiefly concerned with what he considered the preformationism inherent in Albert Kölliker's (1817–1905) developmental application of the Schleiden-Schwann cell theory (See Richmond 2000). Kölliker placed great importance on the nucleus as the essential source of the metabolic and organizational activity in plant and animal development. This made the nucleus and the subsequent cell that supposedly crystallizes around it out to be a preformed source of vital activity. Whereas Huxley was convinced on the evidence of Karl Ernst von Baer's research into *Entwicklungsgeschichte* that development proceeds in an epigenetic fashion, increasingly differentiated form and structure arising out of prior homogenous material. The cell theory, he complained, makes cells—"the primary histological elements"—out to be anatomically and physiologically independent centers of developmental force, so that "the whole organism is the result of the union and combined action of these primarily separate elements" (Huxley 1853, 253–54). As a result, the Schleiden-Schwann cell theory treats a highly integrated animal as if it were a "beehive": "its actions and forces resulting from the separate but harmonious action of all its parts" (254). Schwann had in fact defended the legitimacy of the cell theory by writing that "the failure of growth in the case of any particular cell, when separated from an organized body, is as slight an objection to this theory, as it is an objection against the independent vitality of a bee, that it cannot continue long in existence after being separated from its swarm" (Schwann 1847, 192–93).[30]

Huxley insisted the entire cell theory rested on an incorrect anatomical analysis of plants and animals and consequently led to an incorrect physiological account of development (250). He believed that, far from being the primary agents of development, cells were merely secondary structures, vesicles, or cavities, arising in an otherwise continuous living substance of developing embryos. Because they occurred in what Huxley believed was a continuous mass of developing material, cells are rarely completely separate from one another. The true cause of all organized structure in the plant and animal body, he urged, is more properly ascribed to the vital forces of the component molecules making up the living substance of the embryo (a substance he would later call protoplasm) (Huxley 1853, 261–62). The cells of a plant or animal body, on the other hand, he compared to the "shells scattered in orderly lines along the sea-beach," which indicate only "where the vital tides have been, and how they have acted" (277).

Like Huxley the German botanist Anton de Bary was familiar with the existence of multinucleated masses of protoplasm (syncytia) among lower organisms like the slime molds (myxomycetes) and in animal muscle tissue as well. This raised a difficulty for the conception of the cell as distinct unit or "speck of protoplasm with a nucleus": were these syncytia many cells merged into one? Or are they not cells at all? De Bary also noted that in plants development does not always proceed by the formation of new cells; this is illustrated in many algae such as *Botrydium, Caulerpa, Vaucheria*, where growth in length and differentiation can occur in a single multinucleated mass of protoplasm without any cell division. In fact, even in so-called higher plants, growth and morphogenesis often results in only partial cell cleavage as the addition of the cell plate fails to completely separate the new cell from the old (Kaplan and Hagemann 1991). All this suggested to botanists that growth is more fundamental than cell formation. De Bary famously summarized this belief with the declaration that it is not the cells that create the organism, but the organism that creates cells (De Bary 1879, 222).

Another botanist, Julius von Sachs (1832–97), disputed that the cell "is always an independent living being, which sometimes exists for itself alone, and sometimes 'becomes joined with' others—millions of its like, in order to form a cell-colony, or, as Häckel has named it for the plant particularly, a cell-republic" (Sachs 1887, 73). Like de Bary, Sachs admitted the formation of cells with distinct boundary walls to be a common phenomenon in plant development, but disputed its universality. He considered it to be of secondary significance behind the more general phenomenon of growth exhibited by all organic substance. "According as we keep the one or the other case in view," Sachs wrote, "the cells appear as mere chambers and parts of the growing plant-body, or as independent living organisms from which new plants arise by growth. It depends, therefore, entirely upon our mode of consideration, and upon the point of departure of our consideration, whether we regard the cells as independent so-called elementary organisms, or merely as parts of a multicellular plant" (Sachs, 76–77). Sachs (1892) proposed replacing the structural-morphological concept of the cell with a dynamic and physiological conception, the "energid," which would denote a nucleus and the immediate region of cytoplasm (a term introduced by Kölliker in 1864) directly surrounding it. In Sachs's opinion it was the energid, not the cell, which is responsible for directing the growth and development of plant form.

Botanists were not the only ones taking a more critical attitude toward the cell theory at this time. In 1881 the physician and medical researcher Edmund Montgomery (1835–1911) asked the question, "Are we 'cell-aggregates'?" Definitely not, was his reply. Montgomery appealed to the syncytial nature of muscle tissue and the apparent reticulated fusion of the nervous system into a continuous whole to argue that "structurally viewed, the complex and developed organism, instead of forming an aggregate of 'cells,' forms, on the contrary, an unbroken living substance, or single physiological unit" (Montgomery 1881, 105). Similarly, the anatomist Carl Heitzmann (1836–96) rejected the conception of higher animals as cell aggregates on the evidence that the purportedly separate and distinct cells typically remain connected after cleavage by numerous fine protoplasmic "bridges." This led Heitzmann to conclude that in its microanatomical structure, a higher animal resembles the multinucleated syncytia so common among the Protista. So rather than Man being analogous to a republic or state of amoeba-like cells, Heitzmann concluded: "Man is a complex amoeba"! (Heitzmann 1883, 36).

While the idea that the adult organism consists of a continuous mass of living protoplasm was becoming increasingly popular, there was disagreement about protoplasm's structure: was it essentially fibrillar, foam-like, or granular in construction? Discussion of one particular proponent's argument may be instructive. The physician Louis Elsberg (1836–85) shared Heitzmann's conviction that the various tissues of the body comprised a continuous and connected system, but of what he preferred to call "bioplasson," a particular conception of living protoplasm as composed of ultimate particles he called "plastidules." In a talk before the New York Academy of Sciences, Elsberg claimed to have histological proof of the incorrectness of the cell-doctrine based on his study of the microanatomy of the hyaline cartilage of the human larynx (Elsberg 1881). His slides of prepared cartilage tissue revealed a continuous matrix within which certain "corpuscles" or cells were evident, but he insisted it was not possible to maintain that the structure was entirely cellular. In fact the ground substance and the protein fibers that make up hyaline cartilage are elements of the extracellular matrix, which like other connective tissues, are created by cells known as chondrocytes (Gilbert 2003). Elsberg described these corpuscles as being interspersed throughout the "glass-like" tissue, like "raisins in a cake dough" (Elsberg 1881, 588). But he believed the hyaline cartilage to consist of a continuous network of "fine filaments of living matter" interspersing the tissue and making contact with other differentiated tissues nearby, as if even the raisins in the cake themselves formed a "filigree or framework of raisin-substance".[31] So why, Elsberg wondered, were some people so reluctant to jettison the cell-doctrine? Elsberg quoted John Drysdale, another protoplasm advocate, who opined that "the truth is, this clinging to the mere name of the cell-theory by the Germans seems to rise from a kind of perverted idea of patriotism and pietas toward Schwann and Schleiden" (40).[32] Elsberg demurred from this bit of nationalist criticism, thinking it more likely that the cell had simply become closely associated with developments in histology. But once people came to accept that no part of a higher organism is autonomous from the whole, the body will have to be compared, not to a community or state of cells, but to a "machine, such as a watch or a steam-engine, in which, though there are single parts, no part is at all autonomous, but all combine to make up one individual" (41).

Critics of the cell doctrine framed the issue as a choice between two rival perspectives on anatomy and development. These were contrasted in detail in 1893 by the zoologist C. O. Whitman (1842–1910), who described them simply enough as the "cell and organism standpoints." Whitman appealed to evidence from studies in comparative animal embryology to support the observations of Huxley, de Bary, and Sachs that growth and differentiation of structure are not reliant on the formation of new cells by division (Whitman 1893). The distinct differentiation of structures visible within a single-celled protozoan like *Paramecium* (e.g., an oral groove, numerous vacuoles, and multiple nuclei) was a familiar and increasingly more important example of this point.[33] But the cell concept, Whitman said, was like a ghost that haunts the researcher's observations: "We are so captured with the personality of the cell that we habitually draw a boundary-line around it, and question the testimony of our microscopes when we fail to find such an indication of isolation" (Whitman 1893, 645). Just shortly after this, the embryologist Adam Sedgwick (1854–1918) added another withering critical evaluation of the cell theory in the pages of the *Quarterly Journal of Microscopical Science*. Like Whitman, Sedgwick too considered the cell concept, and the cell theory of development in particular, to be a "fetish" holding "men's minds in an iron bondage." It "blinds men's eyes to the most patent facts," Sedgwick complained and "obstructs the way of real progress in the knowledge of structure"; for the cell is "a kind of phantom which takes different forms in different men's eyes" (Sedgwick 1895, 88–89).

Sedgwick took particular issue with the neuron theory, which proposed that the nervous system is comprised of separate cells (neurons) making contact with one another by the out-growth of cell processes similar in respects to the string-like pseudopods of some amoebae. Drawing on evidence from the development of Elasmobranch embryos (sharks, rays, etc.), Sedgwick claimed that nerves were "so to speak, crystallized" out of a dense cord of multiple nuclei embedded in a continuous mesoblastic reticulum of the embryonic nerve crest. This, he said, should be evident to anyone "not warped by the cellular theory as ordinarily taught" (Sedgwick 1895, 95). Furthermore, his study of the development of the velvet worm (*Peripatus*) revealed the incomplete segmentation of the blastomeres, demonstrating in his estimation that differentiated tissues and organs arise from a multinucleated syncytium, not separate and independent cells. While cell membranes eventually do form around the individual nuclei, the resulting cells are not completely sealed off from one another, for they remain joined by means of the protoplasmic strands mentioned earlier by Montgomery, Heitzmann and others. The existence of these protoplasmic bridges in other cases of animal development proved that cells are not as distinct and isolated as the cell doctrine would make them out to be.

Sedgwick's critique of the cell theory was answered by the zoologist Gilbert Bourne (1861–1933). Bourne stated what he took to be the essential propositions of the cell theory: "The multicellular organism is an aggregate of elementary parts, viz. cells. The elementary parts are independent life units. The harmonious interaction of the independent life units constitutes the organism. Therefore, the multicellular organism is a colony (cell republic according to Häckel)" (Bourne 1896, 163). Of the four propositions, he wished to retain only the first unaltered. The key point of contention, as Bourne saw it, was whether the cells of a multicellular organism could be justifiably considered independent life units. Bourne (165) conceded that cells are only independent *in posse*, since in many cases they are so subordinated to the rest of the organism, on account of the differentiation and specialization that they have undergone, as to have lost their potential for independent activity when separated from the whole.

But in response to Sedgwick's criticism–one frequently made by others as well—that cells cannot really be said to exist if they are connected by continuous strands ("bridges") of protoplasm running between them, Bourne (167–68) asked whether a room ceases to be a room when its door is open. "Is a house to be regarded as one room or composed of separate rooms?" (167). Bourne insisted that like the concept of a room, the concept of the cell does valuable intellectual work for us, and that one should not expect biological concepts to fit perfectly onto the real world. In the end, he concluded "that the cell concept is a valuable expression of our experience of organic life, both morphologically and physiologically, but that in higher organisms cells are . . . not independent life units . . . but a phenomenon so general as to be of the highest significance; they are the constant and definite expression of the formative forces which reside in so high a degree in organic matter" (171). So despite initially coming out strongly in defense of the cell theory, Bourne essentially wound up defending a position not that different from the one outlined by Sachs.

A central component of what Whitman had called the organismal standpoint, was Huxley's earlier suggestion that there is an internal organization in a fertilized egg that subsists beneath the level of the cell; and consequently not all organization and differentiation can be attributed to the composite aggregation of cells. That there exists an internal polarity in the unsegmented egg of the frog was recognized as early as 1834 by Karl Ernst von Baer (Wilson 1900, 378), and the unequal division of cells in the development of many animal embryos spoke further to the importance of a level of organization beneath that of the cell. For this reason, some advocates of the organism standpoint insisted that the fertilized egg should be regarded as an organism rather than a cell. When an egg divides we do not say we now have two egg cells but rather that there are now two blastomeres of an organism in the process of development. Another motivation for the organismal standpoint was the belief that a physical continuity throughout the mass of a developing embryo is necessary to create the harmonious integration of the organism as a whole. Were there in fact a complete separation of the dividing embryonal cells into independent and autonomous units, as the cell standpoint suggested, the coordinated development of this aggregate of elementary organisms into a complete and harmoniously functioning superorganism would be a complete mystery.

At the turn of the century the anatomist Martin Heidenhain offered an illustrative review of the state of affairs in cytology in a two-volume textbook, *Plasma und Zelle* (1907). The title itself suggests a tension between two contrasting viewpoints. Heidenhain explained that it was not his objective to deny completely the legitimacy of the doctrine that plants and animals are aggregates of individual cells (what he more specifically called in its anatomical form the *Bausteintheorie* and in its physiological form the *Lehre vom Zellenstaat*), but to restrict these theories' validity, as they were based on an exaggeration (*Uebertreibung*) of the independence and autonomy of individual cells in plants and animals (Heidenhain 1907, 29). Though sensitive to the need to recognize the interconnectedness of the entire animal and plant body as an integrated system (13), Heidenhain was not ready to give up the cell concept entirely for the more general notion of protoplasm. He offered his own definition of a cell as: "a limitable clump of living matter, with the morphological and physiological character of an elementary individual" (25). To describe a portion of protoplasm as an individual might seem to beg the question, and Heidenhain was quite upfront that the concept of an individual was borrowed (*entlehnt*) from the domain of human life and experience. But in its defense he mentioned five key bits of evidence: (1) the analogy between tissue cells and unicellular plants and animals (protophyta and protozoa); (2) the ability of individual cells to survive outside the animal body in cell culture; (3) the white blood cells of vertebrates, which exist as individuals freely moving about the animal body; (4) the process of cell reproduction, which shows that all cells come from the division of a previously existing individual cell; and lastly (5) the growing recognition that the death of individual tissue cells (*Zellentod*) is a normal part of tissue and organ maintenance (25–26). However, he noted that it must be acknowledged that in the higher plants and animals the independence of the individual cell is typically so subordinated to the organism as a whole that it assumes the role of a mere tool or instrument (*bloßen Werkzeuges*) serving the interests of the entire body (29).

The status of protists as single-cell organisms (the first of Heidenhain's evidence in favor of the cell standpoint) was forcefully challenged by Clifford Dobell (1886–1949), a former student of Sedgwick. That the organisms previously known as the *Infusoria* are composed of a single cell had been promoted in the mid-nineteenth century chiefly by Carl von Siebold (1845), Kölliker (1845), and Haeckel (1866), and constituted a fifth thesis of the cell theory (See Jacobs 1989). Dobell (1911, 285) called for the abolition of the cell theory on the grounds that it was overly simplistic and led to "*Procrustean*" decisions about whether an organism is unicellular or multicellular, when in fact, it may be neither. The Protista, Dobell insisted (272, 276–77, 281), are noncellular or acellular in organization; and this was a fact that only someone blinded by the dogma of the cell theory could fail to recognize.[34] How can it be, he asked, that a complete organism like an amoeba or a paramecium consists of a single cell, if a cell is understood to be a *part* of an organism?[35] A natural response is to say that we need not define a cell as a part of an organism. The problem though, as Dobell saw it, was not that one could not give a clear definition of the cell, but that the definition had been stretched too widely to cover such unlike things as: a whole organism (a protist), a part of an organism (a tissue cell), and a potential whole organism (a fertilized egg). No good could come, he argued, from such verbal confusion.

Dobell's criticisms had a significant influence on other scientists. Consider the following passage from Leonard Doncaster's 1920 textbook on cytology:

> As long as it was generally agreed that all organisms are built up of cells as a house is of bricks, the description of the cell as "a unit of living matter" was not open to any very grave objection; the cell was to the biologist almost what the atom was to the chemist—the smallest portion of living matter capable of an independent existence—and the word had, in appearance at least, a fairly definite meaning. Now, however, when the old idea of discrete and independent cells is almost abandoned, and when distinguished biologists maintain that one whole group of organisms (the Protista) are non-cellular, the word "cell" is beginning to lose its definite and precise significance, and to be used rather as a convenient descriptive term than a fundamental concept of biology (Doncaster 1920, 1).

Just previous to this, the zoologist William Emerson Ritter (1856–1944) published his own lengthy and sustained criticism of the cell standpoint. Arguing passionately that the organism is more fundamental than the cell, Ritter made the insightful point that the tradition of regarding the cell as an elementary organism itself assumes the organismal perspective to be more fundamental (Ritter 1919, 156–57). Why else would it seem at all helpful to think of the cell as an organism if we did not already assume that to speak of an organism assumes a harmonious and integrated principle of organization? In point of fact, Brücke would have very likely agreed with this, for as mentioned earlier, it was not his intention in calling cells "elementary" organisms to assume they are simple or homogenous lumps of protoplasm; they were, rather, elemental (i.e., not further reducible to living parts), or at least Brücke suggested biologists should assume them to be so until further evidence suggested otherwise.

Writing in the early part of the twentieth century Ritter also had to respond to the new developments in tissue culture by researchers such as Ross G. Harrison (1870–1959) and Alexis Carrel (1873–1944). This exciting research revealed that animal cells could survive outside of the body, even crawling about, feeding, and dividing like independent amoebae.[36] While some took this to be solid confirmation that the cells of the animal body are elementary organisms (recall Schwann's remark about the bee subsisting outside of its hive), Ritter had concerns about the reliability of experiments dealing with the behavior of cells and tissues in what were undeniably abnormal conditions (Ritter 1919, 70). For this reason, Ritter preferred to describe the materials at the center of these techniques as *überlebende Gewebe* ("surviving tissues"), to emphasize their unnatural circumstance (168).

Cells, Ritter insisted, should be regarded as "*organs* of the organism just as muscles and glands and hearts and eyes and feet are so regarded" (191; italics in original). Moreover, investigation of the cell by means of the recent methods of biochemistry should, he insisted, lead one to regard the cell as an "organized laboratory," a metaphor that should naturally appeal to an advocate of the organismal conception, since a laboratory, unlike an elementary organism, exists not for its own sake, but only as an instrument for the benefit of a (human) organism. The growing popularity of the CELL IS A CHEMICAL LABORATORY metaphor will be explored more fully in chapter 2.

Still, proponents of the organismal perspective needed to explain why it is that plants and animals so often do assume a cellular organization. This they did by appealing to a principle, credited independently to Rudolf Leuckart (1822–98) and Herbert Spencer (1820–1903), regarding the physical and chemical constraints on how large a volume of protoplasm can effectively grow. Assuming a roughly spherical shape for a mass of protoplasm, any growth in size will see its volume increase more quickly than its surface area, so that the diffusion of gases and other materials between the inner mass and the outer environment required for maintaining homeostatic equilibrium puts a constraint on growing too large. By dividing into more or less distinct cell units, a mass of living protoplasm can overcome this limitation on growth. In this way the cellular organization of plants and animals would be a secondary phenomenon, as Julius Sachs had said, behind the more essential phenomenon of growth.

This account put considerable wind in the sails of people like the marine biologist and philosopher of biology E. S. Russell (1887–1954), who released his pro-organicist and anti-reductionist review of theories of embryology and animal development in 1930. "We shall see," he wrote, "that the cell-theory has much exaggerated the importance of the cell, and obscured the importance of the organism" (Russell 1930, 197). Russell was in general agreement with Ritter (Russell, 177). The "classical cell-theory" "asserts that all multicellular organisms are built up of fundamental units, the cells, as a house is built of bricks or stones" (Russell 1930, 234); whereas, he argues by analogy, "the cells taken separately no more constitute the organism than words or letters by themselves make a sentence" (235). Like the complete sentence, the organismal context in which the cells of multicellular plants and animals develop and function is essential for understanding the proper significance of the cell parts.

In the decades around the turn of the twentieth century, the legitimacy of the cell as the proper unit of analysis had come under attack not only from above but from below as well (Dröscher 2002). A menagerie of micromeristic units were said to be more fundamental living units than the cell itself. A partial list of these includes Herbert Spencer's physiological units (1864), Carl Nägeli's micelles (1867), Darwin's gemmules (1868), Louis Elsberg's plastidules (1874), Richard Altmann's bioblasts (1890), Sach's energids (1892), and August Weismann's biophores (1892). Some of these would prove to be nonliving subcellular organelles, some (like Altmann's mitochondrial bioblasts) would turn out to be formerly independent cells now living as endosymbionts in eukaryotic cells, while others (like Darwin's gemmules and Weismann's biophores) were purely hypothetical. But the resolution of the fine microstructure of the cell and its cytoplasm would require tools and techniques not fully developed till the rise of molecular biology beginning in the mid-twentieth century.[37]

8. The redoubtable cell

In spite of this battery of criticism, the cell proved to be a redoubtable opponent.[38] What had taken most of a century to establish, that the cell was not merely an artifact of inferior optical lenses but a real biological entity and a legitimate concept of scientific analysis and explanation, was not easily abandoned. Hence its conspicuous place at the center of what is recognized to be one of the most celebrated textbooks of cytology at the turn of the twentieth century, *The Cell in Development and Inheritance* (1st edition 1896; 2nd 1900; 3rd 1925) by Edmund Beecher Wilson (1856–1939) (Maienschein 1991a; Dröscher 2002). Wilson's approach in cytology was strongly informed by the morphological tradition, although his text focused on both cell structure and function. The second and third editions indicated the changing nature of cytological studies, as reliance on physicochemical approaches for the investigation of the cell's role in development and heredity in particular became more common.

Wilson by no means discounted criticisms of the cell concept, admitting that, "The term 'cell' is a biological misnomer; for whatever the living cell is, it is not, as the word implies, a hollow chamber surrounded by solid walls" (Wilson 1896, 13).[39] Wilson adopted the Leydig and Max Schultze definition of a cell as "a mass of protoplasm containing a nucleus," with the added remark that "both nucleus and protoplasm arise through the division of the corresponding elements of a preexisting cell" (1900, 19). However, he noted, as many others had before him, the failure of attempts to replace the word with a more suitable one, so that it seemed wise to accept that the cell is here to stay (Wilson 1896, 14; and similarly in 2nd ed. 1900, 19). And yet Wilson did lament that Sachs's word "energid" had not become more popular (1896, 14, n. 1).

As for the cell theory, in the first and second editions, Wilson defended its principal claims by appealing to several well-known achievements: the establishment of sperm cells as motile single cells by Kölliker in 1841; the amoeboid movement of white blood cells throughout the blood in 1844 by Thomas Wharton Jones (1808–91), and Remak's description in the mid-1850s of the development of the animal embryo by means of a series of binary cell divisions from the egg or ovum, which was itself recognized as a single cell.

But while conceding that the autonomy of tissue cells had been exaggerated (Wilson 1896, 41), in the third edition Wilson defended the "cell-state theory" on the basis of the valuable advances it had made possible (Wilson 1925, 101–2). Among these he counted "a revolution in the prevailing views of vital action" and a great impetus to the fields of physiology, pathology, and morphology (Wilson 1925, 5). While admittedly requiring some qualification, "especially as applied to the phenomena of growth," he wrote, "the conviction of its essential truth has survived all criticism, and as measured by its continued fruitfulness, it still stands among the most important generalizations of modern biology" (5). Evidence that cell autonomy is typically subordinated to the interest of the whole organism did not undermine what he described as "our fundamental conception of the cell-state" (103).

Wilson discussed experiments by Hans Driesch (1867–1941) with sea urchin embryos that had revealed that development of complete larvae could withstand the destruction or removal of individual blastomeres. Although such results admittedly seem opposed to "Schwann's conception of the multicellular organism as a composite or mosaic" (Wilson 1925, 1031), Wilson insisted that other recent experiments counted in its favor. Here Wilson had two sorts of experiments in mind, each also dealing with the rearrangement and manipulation of animal embryos and the cells of marine invertebrates (Wilson 1925, 1030–32). The first involved experiments carried out by H.V. Wilson (1863–1939) on sponges. Wilson (no relation) forced sponges to disaggregate into individual cells by cutting them into small pieces and then forcing them through fine cloth. After a time, the individual amoeboid cells began to reaggregate and to reassemble into new sponge organisms (Wilson 1907). The other experiments concerned experimental manipulations of embryos showing that pieces of embryos transplanted to new locations in the embryo continue to develop as they would have if left in their original environment, and that blastomeres removed from the embryos of some organisms altogether will continue to develop the specific part they would normally contribute to the complete embryo. E. B. Wilson emphasized that such experiments show that cells can retain a high degree of independence despite also being undeniable parts of an organism.[40]

Wilson's choice of figures to illustrate cells reflects his attempt to maintain a balance between the organismal and the cell standpoints. For instance, his first figure depicts a number of tissue cells from the epidermis of a larval salamander all in close contact with one another, as does his second figure (added in the second edition) showing the cells from the growing root-tip of an onion. His third figure, however, is of an isolated and free-living amoeba (*Amoeba Proteus*, to be specific); and in what represents a more innovative move, he used in all three editions one of the first highly abstract and schematized diagrams of a "typical" cell, which is most significantly presented in complete isolation (see figure 1.6).[41]

In the midst of these disputes over the cell's status, the embryologist and philosopher of biology J. H. Woodger (1894–1981) attempted to bring some conceptual clarity in his book *Biological Principles* (Woodger 1929). Woodger suggested that the disagreement resulted from the disputants having confused three separate notions of what was meant by a cell. In one sense to talk about cells is to refer to a particular type of biological organization commonly found in many types of organisms.

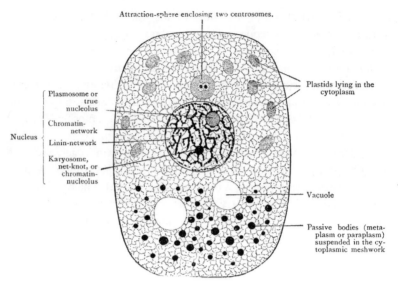

Attraction-sphere enclosing two centrosomes.

Plastids lying in the cytoplasm

Plasmosome or true nucleolus

Chromatin-network

Nucleus

Linin-network

Karyosome, net-knot, or chromatin-nucleolus

Vacuole

Passive bodies (meta-plasm or paraplasm) suspended in the cy-toplasmic meshwork

Fig. 6.— Diagram of a cell. Its basis consists of a meshwork containing numerous minute granules (*microsomes*) and traversing a transparent ground-substance.

FIGURE 1.6 Diagram of a "typical" cell. Note its complete isolation. (From Wilson 1925.)

Some complete organisms, like the protozoa, spend their entire life history with only the cell-type organization, while others, e.g., the metazoa, have parts some of which can be characterized as displaying the cell-type of organization, but others having an organization above the cell-type, namely the tissue and organ type (Woodger 1929, 296ff.). This sense of what one means by a cell is obviously a conceptual abstraction: it is a concept not a material thing or entity. But in a second sense, people talk of a cell when referring to a particular entity observed through a microscope: "This cell here." Woodger believed, however, that people were mistaken if they assumed this concrete and particular instance of observation singled out an objectively real entity. What was concrete and particular in such a scenario, he argued, was a limited spatiotemporal event (the influence of Whitehead's process philosophy is evident here), whereby we single out what Woodger called a "perceptual object" (159–60), which he understood as also being conceptual in nature. The identification of the perceptual object as a cell brings us to the third sense of what is meant by a cell. This, Woodger explains, involves the abstract conceptualization or interpretation of a unique spatiotemporal and particular segment of the world (the latter of which he called the "primary realm"), as an instance or instantiation of a general type or universal. While the naive realist believes he has witnessed a concrete cell, in fact he has *conceptually interpreted* a concrete event *as a cell*. In essence, Woodger is making the point that observation is theory-laden.[42] What is "given" to us in perception is an indeterminate and undistinguished mass of sense impressions (the primary realm), but what we observe or see is informed by our education and training (and Woodger might have added our language or vocabulary to the list). "We sense the primary realm but we perceive the object" (137), or, put in Kantian terms, "There is no perception without conception."

Both parties in the dispute, Woodger felt, needed to recognize that whether we talk of "the cell" or "a cell" we are dealing with abstractions. While critics of the cell concept seemed to believe they had dealt a deadly blow by pointing out that the cell concept did not exhaust every feature of every organism, its defenders (he mentions E. B. Wilson) seemed to believe that by pointing to real instances of cells they were meeting their critics' charges with "concrete facts" (160). The value of any abstract concept, Woodger emphasized, is its ability to highlight certain features shared by many different things, but it is no failure on the part of a concept that it not exhaustively cover every detail of every thing to which it is applied. So the proposition that "some organisms are such that they can be analyzed into parts all of which are characterized by the cell-type of organization" he admitted is "true and important as long as it is not forgotten that (1) there are parts having an organization above the level of the cell-type, and (2) there are parts which are not cell parts, and not analysable into cells, although these are not living," by which he had subcellular organelles in mind (Woodger 1929, 298).

Despite the value of the cell concept as a useful abstraction, Woodger insisted that biologists recognize that many organisms of a higher type cannot be analyzed completely as aggregates of cell-units, for not to do so would be to ignore the fact that there are other types of organization in which the individual cells are "transcended," as he put it, so as to create extra- or supracellular parts or systems (308, 356). Through their interrelations, cells form "territories" (an idea expounded earlier by Goodsir and Virchow), which separate individual cells into distinct regions with peculiar properties, e.g., the germ layers of the developing embryo. "With the attainment of such parts," Woodger writes, "we now have the possibilities of new types of relations between one *cellular* part and another" (356). Woodger here points to the important transition from individual cell behavior to cell-group behavior, a topic sometimes known as "cell sociology," discussed at length in chapter 3.

The cell concept and the cell theory had clearly suffered a kind of demotion in the estimation of many cytologists. In his influential *Text-*

book of Experimental Cytology, the zoologist James Gray (later Sir James) admitted the obvious significance of the cell *as a unit of structure*: "Throughout life the cell remains the unit of organisation just as the bricks remain the units of which a house is built" (Gray 1931, 1). But *as a physiological unit* its ultimate significance was less clear. Gray mentioned that nearly all experimental biologists at that time rejected the view of the metazoan animal as a colony of cells, and he cautioned that biologists must be "careful to avoid any tacit assumption that the cell is a natural, or even legitimate, unit of life and function" (2). Invoking the Spencer-Leuckart explanation for the organism's frequent but by no means universal need to break up its living protoplasm into cell-units, Gray remarked that the cell "is merely the unit of mechanical stability," whereas "the real unit of life must be of a protoplasmic nature irrespective of whether it is subdivided to form a mechanically stable system or not: in other words, cellular structure is not in itself of primary significance" (3). On the other hand, he admitted that the cell was often a convenient unit for physiological analysis (4), yet he made it clear that it was only "as a convenient unit of functional activity that the cell will be regarded in this book" (5).

It appeared then that the cell had fallen a long way from fundamental unit of life to a merely "convenient unit of functional activity." It was in this climate of increasing skepticism that the Oxford cytologist John Baker (1900–84) embarked on his extensive review of the history of the cell concept and the evidence on which the cell theory stood (Baker 1948–52). Baker began by noting that the cell-theory comprised a complex of seven different propositions in total, any or all of which may be the focus of attack by critics (Baker 1948, 103, 105). These were as follows:

1. Most organisms contain or consist of a large number of microscopical bodies called "cells," which, in the less differentiated tissues, tend to be polyhedral or nearly spherical.

2. Cells have certain definable characters. These characters show that cells (a) are all of essentially the same nature and (b) are *units* of structure.

3. Cells always arise, directly or indirectly, from preexisting cells, usually by binary fission.

4. Cells sometimes become transformed into bodies no longer possessing all the characters of cells. Cells (together with these transformed cells, if present) are the living parts of organisms: that is, the parts to which the synthesis of new

material is due. Cellular organisms consist of nothing except cells, transformed cells, and material extruded by cells and by transformed cells (except that in some cases water, with its dissolved substances, is taken directly from the environment into the coelom or other intercellular spaces).

5. Cells are to some extent individuals, and there are therefore two grades of individuality in most organisms: that of the cells, and that of the organism as a whole.

6. Each cell of a many-celled organism corresponds in certain respects to the whole body of a simple protist.

7. Many-celled plants and animals probably originated by the adherence of protist individuals after division (Baker 1948, 105–6)

In the end, Baker defended weakened and qualified versions of each of these theses. The existence of syncytial organisms, he decided, "constitutes an exception to the cell-theory" (Baker 1952, 157), so that the cell theory is true only of "most" organisms (Baker 1948, 105, 123). Cells are "to some extent" individuals; each cell of a many-celled organism corresponds "in certain respects" to the entire body of a protist organism, and multicellular plants and animals "probably" originated through the failure of protist individuals to separate after division (Baker 1948, 106). The cell theory, according to Baker's appraisal of the evidence, is not a universally valid truth but an important generalization of significant pedagogical utility.

The effect of all this critical discussion was not a complete abandonment of the cell theory or the cell concept, but a more nuanced and measured appreciation of their strengths and weaknesses. This is nicely illustrated in a work remarkable for its combination of technical science and philosophical sophistication as much for its provocative title, *The Organization of Cells and Other Organisms* (1960), by the zoologist and scholar of Asian music, Laurence Picken (1909–2007). Picken stressed the importance of the history of a scientific discipline for its future progress, and, in particular, awareness of how its key concepts have changed over time. To ignore the changing character of concepts such as "the cell" is to ignore the "provisional character of any scientific generalization" (Picken 1960, xxxiv):

A knowledge of the history of changing ideas in the past, far from being a luxury, is essential as a means of accustoming us to, and preparing us for, the possibility of ideas changing in the present and the future; and it may make us the more ready to experiment in enlarging or revising our concepts; it will make us self-conscious about our habits of scientific speech; it will tend to make us aware of the gyves and manacles of language which set limits to our powers of observation.

Picken stressed the importance of hypothesis as stimulus to further inquiry, even if based on "insufficient or misinterpreted data"; and because of this essential stimulating character it is not to be suppressed, "however daring." "In a pragmatic biology," he writes, "it might be accepted that a theory—or let us make it simply a concept—is superfluous unless it leads to action" (xxxv). Science is not the purely logical exercise it is sometimes made out to be, and scientists "have a way of being right, though their premises are wrong and their logic faulty" (xxxvi). To the objection that there is danger in entertaining insufficiently established hypotheses Picken responded, "There is danger to no one, and those only will be disconcerted who believe that science is Absolute Truth. Of that, science has no knowledge." Picken concluded his introductory remarks with a defense of the importance of the "catalytic activity of ideas" for the process of scientific discovery (xxxvii).

The controversy over the cell theory was complicated, as Baker had shown in his review, by the fact that there is no single theory but a set of heterogeneous propositions: some taken by many biologists to be matters of fact, some assertions, the truth of which depended on how one interpreted their key terms, and some that are untestable hypotheses (Picken 1960, 2). There were, however, two main issues, as Picken saw it: one, the extent to which cells are discrete units rather than continuous segments of protoplasm organized around one or more nuclei (Sachs's *energid*); and two, whether the Protista are properly characterized as unicellular organisms and so comparable to the tissue cells of multicellular plants and animals. With respect to the first, much hung on how one understood the notion of cells being isolated one from another. Writing from the vantage point of the mid-twentieth century, Picken exclaimed, "We are now so aware of the constant exchanges between cell and cell, and cell and environment, that the idea of a cell as a unit *isolated* by its membrane is quite foreign" (3). I will discuss in detail in chapter 3 the developments that made such an idea so unlikely.

As for the second, he remarks that Dobell's criticisms were justly made toward an over-simplification and overgeneralization, and that Dobell was largely motivated by the desire to have the Protista recognized as complex modern organisms in their own right rather than as stand-ins for primitive ancestors in recapitulationist theorizing. Picken believed that Woodger had pointed to the resolution of this particular difficulty by noting that "the cell" was a highly abstract concept of minimal content, denoting a particular type of organization that was applicable to many Protozoa; while talk of "a cell" from a metazoan seen under a microscope involved a concept of a lower grade of abstraction (a perceptual object) (3–4). Keeping this distinction in mind, he suggested, ought to prevent any further controversy.

One important question remained, namely, whether the cell concept could legitimately be extended to cover all forms of life and thereby recognized as *the* unit of life. This required resolving the question whether viruses and bacteria were of the cell-type of organization. To call a bacterium a cell would require extending the current definitions of "nucleus" and "cytoplasm." In order to bring the prokaryotes—microorganisms lacking a nucleus—into the cell theory tent, the cell concept had once again to be revised by substituting nucleic material (chromatin or DNA) for a nucleus as an essential feature.[43] Picken recommended (22) describing them as "protocells" and to adopt the term "eucells" to refer to the cells of the Protozoa, Metazoa, and Metaphyta (collectively the Eukaryota in present jargon). This left the viruses, which though highly organized, lack the independent metabolic capacities of standard living organisms. But because the independence of living organisms is always qualified and never absolute, this he felt offered no clear reason for not recognizing them as organisms; and given that other living organisms are known to enter into and out of dormant nonliving states (e.g., fungal spores, tardigrades, and rotifers), viruses did not seem so extraordinary in this regard. Their similarities to the genetic elements of eucells suggested they may be fragments of chromosomes from identifiable genomes, but until more evidence was available it would be best, he decided, to recognize viruses as a distinct hierarchy of organisms. They certainly were not cells according to his criteria (for they lacked a nucleus and cytoplasm). Picken's decision to extend the concept of an organism in such a way that all cells are organisms but not all organisms are cells explained his choice of title, *The Organization of Cells and Other Organisms*. Of greater interest is his further decision that "if the cell is *the* unit of life they [viruses] cannot *ipso facto* be living" (24). This bit of reasoning illustrates with exceptional clarity how the cell-theory could move beyond the status of a mere generalization or empirical law and function as an almost irrefutable tautology (Reynolds 2010).

Picken's text is also illustrative of a movement in the mid-twentieth century of what might be called "the return of the cell-organism." For despite the intensive criticism levied against the idea that an animal or plant cell is a distinct autonomous unit, an elementary organism, many biologists continued to find this an attractive and fertile way of thinking and talking about cells.[44] This is one sense in which the cell has been re-doubtable: it has proven very resilient to the criticisms of its opponents. And yet those attacks and criticisms have not entirely let up even to this day. That its legitimacy has remained subject to doubt constitutes the other sense in which the cell has proven to be "re-doubtable."

Many biologists have continued to object to the supposed independence and autonomy on which rests its claim as a, let alone *the*, unit of life. Metaphors of building stones and citizen-individuals have continued to cast their shadows from behind the screen against which the major conceptual actors have played out their parts. For instance, in the late 1960s the cancer biologist Sir David Smithers (1908–95) felt compelled to challenge the assumption that the mammalian cell is a self-contained unit, insisting that "an organism cannot be broken down into its component parts and rebuilt step by step as a house may be" (Smithers 1969, 778). The title of his letter to the editor of the *British Medical Journal*—"No Cell is an Island"—highlights, in a way that manages to be both metaphorical and literally true, that cells are no more immune to the influence of their "peers" or the broader external environment than are humans living in society.

The upshot of all this criticism of the cell's supposed autonomy is that the cell is now meant to be understood as a unit without necessarily being unitary, i.e., isolated and entirely autonomous. And this I think explains its perpetual attraction as a conceptual tool of analysis. Even today some writers, albeit primarily for didactic purposes, find the metaphor A CELL IS A LEGO BLOCK helpful for thinking about how multicellular organisms are composed from simpler units.[45] Even some of the cell's current critics are not entirely opposed to the notion that some kind of repeatable unit or pattern can be identified by which complex organisms may be at least conceptually decomposed and understood. The plant biologists Baluška, Volkmann, and Barlow (2004) are critical of the thesis that plants are composed of discrete cell-units separated by solid walls (for they note how plant cells are connected by plasmodesmata and share a common plasma membrane), but in arguing that higher plants have a "supra-cellular organization," their proposal consists in a replacement of the cell concept with a modernized version of Sachs's energid concept, which they call the "Cell Body" (consisting of the nucleus and its associated microtubules). According to their proposal, the cell as typically conceived is not the smallest independent unit of life, what they call the Cell Body is. There is irony here in that the cell theory was first motivated by the study of plants, but it is plant biologists who now most vociferously reject the cell standpoint as an adequate account of plant growth, anatomy, and physiology.

Other cell critics have proposed a return to protoplasm as a more adequate concept. The physiologists G. Rickey Welch and James Clegg (Welch and Clegg 2010; 2012) argue that protoplasm can provide a central concept around which current talk of "systems biology" can be better organized than continued talk of the cell, which is after all "simply the 'housing' of life's basic physical, chemical and genetic processes" (2012, 644). Continued emphasis of the cell over protoplasm, they say, reflects a conviction that form dictates function rather than function driving form (the protoplasm theory). The authors note that the dominance of the cell theory over protoplasm theory was likely because the cell marked out a "concrete and easily identifiable physical entity, readily visible with light microscopy," which proved to be a "more objectifiable and safer philosophical construct" (644). I would add that the attraction of the cell over protoplasm is that whereas protoplasm denoted a substance—and a highly heterogeneous one, as it turned out, the cell marks an individual unit, an agent of sorts about which we can tell a story and with which we humans can identify. To use a homey analogy, suppose we are in the kitchen baking. Now we can choose to talk about cookies or about dough, and while it is true that cookies are made from the fundamental dough-stuff, we would miss something important—for certain types of investigations and questions—were we always to prefer talking of dough/protoplasm over cookies/cells. But which is the proper perspective will depend ultimately on the sort of questions being asked.[46] Ultimately, however, Welch and Clegg may be confusing the name (cell) with its current cognitive content, for what we now mean by a cell is almost entirely independent of its original morphological connotation.

As the structural contents of the cell (centrosome, chromosomes, mitochondria, Golgi apparatus, etc.) and their properties swelled in number and in detail in the late nineteenth and early twentieth century, the original analogical feature after which the cell had been named faded from consciousness, so that today the cell simply connotes the minimal unit of life, with all its special complexities and features. Few people today even recognize the cell to be a metaphor. When this happens, when a metaphor ceases to be recognized *as* a metaphor, it is sometimes said to have become literal, or to be a dead metaphor. Whether either of these ways of talking about such a transition is fully adequate will be discussed below.

To sum up the developments covered so far, step one in the history of the cell theory (covering roughly the period 1840–60s) was to establish the cell as a legitimate unit of analysis for anatomists, physiologists, embryologists, etc. This required emphasizing its existence as a discrete and autonomous entity, not just as a subordinate part of plants and animals, but as an organism in its own right and as a whole. This required an attitude William Bechtel (2010) has described as taking the cell as an object of study, not just a locus of study, as it later became in step two. This second step (roughly from the 1870s to 1910 or so) involved identifying the cell's own parts, its structures, and their properties. By the end of the nineteenth century, cytology had become increasingly experimental and less morphological. Cytology became *Zellforschung* (Dröscher 2002), with special attention being paid to the events and mechanisms of fertilization and division, the chromosomal theory of heredity, its great achievement, joined with Mendel's theory of inheritance to create modern genetics. This was a period of investigation driven by new techniques of staining and fixing organic materials for microscopical inspection, of sectioning tissues to reveal their cellular organization and the cell's own internal organization. A third period of research into cell physiology running roughly from 1900–50 relied more heavily on biochemical techniques and concepts.

Conclusion

While the cell began as one specific metaphor, an empty chamber into which living juices or protoplasm might be contained, what people understood talk of the cell to mean quickly changed, and in several different ways. When the existence of a clearly visible wall or membrane faded in importance, understanding of what was meant by the cell shifted more heavily to the notion of an elementary organism, like an Amoeba, which illustrated the essential features of the new protoplasmic cell concept. Many eighteenth- and early nineteenth-century thinkers, influenced by romantic *Naturphilosophie*, were searching for some fundamental archetype by which to organize and to unite all the diversity of living forms on earth. With the microscope to assist the search for this master-form, investigators began to mentally dissect plant and animal bodies into various smaller units: fibers, globules, vesicles, and cells. The original cell concept of a hollow chamber or vesicle provided one possible search image with which observers could scan through various plant and animal tissues looking for signs of a concurrence in structure. Other units (e.g., globules or vesicles) might have worked as well in principle, but with its emphasis on a distinct and readily identifiable wall enveloping a nucleus, the cell may have had an advantage over the others at this early stage of microanatomical and physiological thought. More importantly, because its chief promoters (Schleiden and Schwann) had also spoken of the cell as an organism, it was taken up by those like Virchow and Haeckel, who developed it into the theory of the cell-state, which guided attempts to understand the relationship between the cells and the larger organism by means of the analogy with how the members of a modern state exist as separate individuals while working together toward a common goal, and in doing so dividing among themselves the various forms of work to be done, and being arranged into various professions or classes. The result was a powerful unifying generalization about life in all its complex and diverse forms.

The philosopher and theoretical biologist Woodger explained its appeal quite well by comparing the biologist's situation to that of someone observing for the first time a wallpaper with a complex and intricate design. The mind almost instinctively looks to identify some simpler pattern, which through its repetition the larger impression is created. When such a pattern is found, we feel we understand the complex whole (Woodger 1929, 137). So despite its initial and lingering inadequacies, the cell concept (or rather, concepts, since the understanding of the cell, as we've seen, was under almost constant development and revision), because it fulfilled this intellectual craving for simplicity beneath a more immediate confusion of diversity, continued to enjoy a wide popularity among scientists. The repeated criticisms of those whose own areas of expertise tended to highlight the points at which the cell theory is a rather poorer fit, remind us of Whitehead's suggestion that the scientist's guiding objective should be to "seek simplicity and distrust it" (Whitehead 1955, 163).

As we shall see in chapter 3, social analogies and metaphors allow us to understand how cells can be thought of as physically distinct entities without being physiologically or phenotypically independent of outside influence from the environment or other cells. While from an anatomical perspective cells may be abstractly thought of as building stones or Lego blocks, from a developmental perspective they share more relevant similarities to human individuals, who grow and develop in a social context of a great degree of mutual communication. There is no such communication between stones or blocks. But humans can be at once distinct individuals independent in one sense (i.e., spatiotemporally) and not independent in another, since their appearance and behavior is affected by the influence of other individuals with whom they interact. That we refer to the products of cell-cleavage as "daughter cells," that we speak of a cell's "fate" (its *Schicksal* in German), illustrates how strong our impulse is to cast cells in a narrative light appropriate to agents, how captivated we are, to use Whitman's phrase, by the "personality of the cell." The continuity yearned for by the protoplasmists and those adopting the organismal standpoint is accounted for within the cell standpoint by recognizing this social aspect of cells, how their communication by means of molecular and mechanical signals literally shapes and influences their morphology and behavior. This will be discussed in detail in chapter 4.

This chapter has been predominantly concerned with the question "What is a cell?," many of the attempted answers to which have focused on morphological considerations. This may be natural as a *cell* in its original meaning is after all a morphological or structural concept. In the next chapter we will look at how biological cells were conceived and investigated from a more physiological and biochemical perspective.

2 Biochemical Conceptions of the Cell: From bag of enzymes to chemical factory

> Analogies are dangerous concepts, but it looks as though the only real conception of a living cell as a dynamic unit is provided by comparison with suitable types of inanimate machines.
>
> **James Gray (1931, 30)**

1. Introduction

While some anatomists and developmental biologists (embryologists) looked to the cell as the unit by which more complex plants and animals are constructed, as a house is built from bricks or stones, and those interested in how multicellular organisms evolved from more ancient forms could look upon the cell as an elementary organism, physiologists and biochemists were interested in how these elementary organisms work. And as the attractiveness of the Haeckelian phylogenetic program of attempting to retrace the evolution of modern plants and animals from more primitive protist ancestors faded at the end of the nineteenth century, attention turned to questions more functional than hypothetical, and even in embryology methods became more experimental, as illustrated by the shift from the largely descriptive approach of *Entwicklungs-geschichte* to the more interventionist techniques of *Entwicklungs-mechanik* (Allen 2007; Churchill 2007). For scientists interested in how cells manage to grow, to move, to divide, and to achieve all the other special tasks they perform as differentiated tissue-cell types, there was little suggestive guidance to be had from saying the fundamental unit of life is a building stone (which is completely inert) or an elementary organism (when the task at hand is precisely to figure out how organisms work in the first place). For this reason, scientists working on these sorts of physiological problems tended to think metaphorically of the cell as a chemical laboratory, factory, battery, or an electric motor. By invoking these sorts of analogies to dynamic and work-performing systems, scientists could approach the investigation of cellular function from within a perspective conducive to application of the principles of physics and chemistry, and the emerging theory of energetics.[1]

The chief background metaphor of this chapter, and that the most general, will be CELLS ARE ARTIFACTS. More specific than this is the background metaphor CELLS ARE MACHINES; and even more specific will be the metaphors CELLS ARE LABORATORIES and CELLS ARE FACTORIES, both which emphasize, or more correctly, hypothesized, an internal organization in the cell, an orderly arrangement of parts directed toward the purposive completion of some type of work or product by means of systematic stages. In discussing how this conception of the cell interior as a highly ordered space came about we will also mention the rival metaphor that CELLS ARE BAGS OF ENZYMES.

A machine is a type of technology, while laboratories and factories employ technologies and machines of various kinds. These reductionist and mechanistic metaphors both promote and assist experimental investigations into the causal working of cells, which, as Bechtel (2006) has shown, involves a specific form of scientific explanation reliant on the notion of a *mechanism*. Although it was already present to a lesser degree in the nineteenth century, it became absolutely commonplace in the twentieth century to think of the cell as a kind of machine. And what do we do to machines? We disassemble them, look at the parts, see how these parts fit together, and ultimately look for ways to make the machine work more efficiently or adapt it for application to other tasks we have an interest in seeing performed. Any time we think of something as a machine, then, we are making the assumption that the thing *has* discrete parts, and that the properties of the individual parts explain the properties of the whole. Less obviously, however, when we describe something as a machine we also implicitly suggest that it is only natural and hence proper to take it apart and attempt to "improve" it for our own purposes. For metaphors are not only *descriptive* in nature, they can also be *prescriptive*. Machine metaphors are essentially tools of reductionist method and interventionist projects that can ultimately result in fundamental changes to the natural properties and behavior of the thing studied. As we will see in this chapter, the cell's journey from metaphorical to literal factory in late twentieth-century biotechnology reveals that metaphors are not only capable of producing conceptual change in the way we see or think of things, they also have the power to bring about concrete material change in the very nature of the objects to which they are applied. Metaphors, therefore, are not just epistemological tools, they are also tools of ontological significance; they might even be called a form of experimental technology.

2. The cell as a chemical laboratory or factory

It is not at all unusual today to run across some version of the following description of what scientists understand a cell to be:

> **All living things are made from cells, the chemical factories of life.** Cells act as chemical factories, taking in materials from the environment, processing them, and producing "finished goods" to be used for the cell's own maintenance and for that of the larger organism of which they may be part. In a complex cell, materials are taken in through specialized receptors ("loading docks"), processed by chemical reactions governed by a central information system ("the front office"), carried around to various locations ("assembly lines") as the work progresses, and finally sent back via those same receptors into the larger organism. The cell is a highly organized, busy place, whose many different parts must work together to keep the whole functioning. (Hazen and Trefil 2009, 252)

The venue in which a metaphor occurs is important, for the intended audience of a piece of writing will have crucial significance for the choice of language, metaphors, and analogies adopted by the author. The quotation above occurs in a book of popular science whose specific purpose is science education for nonscientists. One might be inclined, therefore, to wave off such explicitly metaphorical language as of negligible importance for the actual conduct of scientific investigation and scientific explanation. But we will see, if it hasn't already been made clear in the first chapter, that scientists rely on metaphors for more than just communicating with nonscientists.

Metaphors are said to play three chief roles in science (e.g., Bradie 1999): (1) a rhetorical or communicative role (which would include pedagogical purposes); (2) a heuristic function involving the creation of novel ideas or hypotheses; and (3) a cognitive or theoretical function whereby metaphors are said to help to provide explanations of phenomena.[2] Corresponding only roughly with these three functions are three chief types or genres of scientific writing: popular science writing, original research papers, and review papers.[3] Pedagogical or didactic metaphors are obviously expected to be found in the first genre, though they may also appear in the other two if the author is hoping to educate other scientists unfamiliar with a particular scientific topic (as frequently happens in review articles). Metaphors, according to some, may play an important role in *the context of discovery*, but not in *the context of justification*, where the real cognitive and explanatory work of science is done. If, as many seem to suppose, explanation requires objective, true knowledge, then metaphors cannot be explanatory. Others, notably Hesse (1966), Bradie (1998; 1999), Lakoff and Johnson (2003), Brown (2003), and Ruse (2005), have disputed this, and the widespread occurrence of key *theory-constitutive* metaphors illustrated in this book will, I hope, help to make the importance of metaphors for the creation of scientific explanation more widely recognized.[4] But I leave detailed discussion of these topics for later (see chapters 5 and 6), and return to the metaphors that the cell is a chemical laboratory or factory.

We saw in the last chapter how the critic of the cell standpoint William Ritter had claimed that biochemical studies lead one to regard the cell as an organized laboratory (Ritter 1919, 191). This metaphor naturally appealed to Ritter since, unlike an elementary organism, a laboratory does not run itself or exist for its own sake; rather, it is an instrument used by the organism for its own ends. In the early twentieth century, when biochemical investigations of the cell came increasingly to dominate the attention of scientists, the chemical laboratory or factory metaphor of the cell began to rise in popularity.

Descriptions of the cell as a chemical laboratory can be traced back to the mid-nineteenth century. François-Vincent Raspail (1794–1878), an early student of cell chemistry, wrote in 1843 that "each cell selects from the surrounding milieu, taking only what it needs. Cells have varied means of choice, resulting in different proportions of water, carbon and bases which enter into the composition of their walls . . . A cell is therefore a kind of laboratory within which all tissues organize and grow."[5] Others, such as the zoologist Henri Milne-Edwards (1800–85), writing in 1851, compared the body of a living organism to a workshop ("un atelier") (Milne-Edwards 1851, 35).[6] A year later Franz Unger referred to the plant cell as a "powerful chemical workshop" (*mächtige chemische Werkstätte*) (Unger 1852, 23) and cells as "tiny chemical laboratories" (*kleinen chemische Laboratorien*). And in 1858 Virchow remarked that "starch is transformed into sugar in the plant and animal just as it is in a factory" (*Fabrik*) (Virchow 1858, 107). Such comparisons were likely aided by the chemist Friedrich Wöhler's (1800–82) successful laboratory synthesis in 1824 of the organic chemical urea, previously believed to be exclusively produced by living organisms. Wöhler showed that organic compounds could be manufactured in the laboratory while working with the chemist Justus Liebig (1803–73), whose commercial synthesis of organic dyes for the textile industry helped make the German economy one of the strongest in the late nineteenth century. The nineteenth-century historian Theodore Merz noted that "physiology and economics joined hands" in the Victorian period through the concepts of the "autonomy of the cell" and the "physiological division of labour," and did so chiefly through the influence of Liebig (Merz 1965, 395–96, 415). Merz also wrote that "Liebig looked upon nature on the large and on the small as an economy . . . Through Liebig chemistry entered into close alliance with political economy" (395). The shift from the laboratory to the factory metaphor may in fact mirror the socioeconomic changes attendant with the rise of industrial systems of production in Europe. The physiologist and biochemist Claude Bernard (1813–78) described the structure and function of animal organs in this way: "In the living body these organs are like the factories [*les manufactures*] or the industrial establishments in an advanced society which provide the various members of this society with the means of clothing, heating, feeding, and lighting themselves" (Bernard 1885, 358).

Descriptions in terms of metaphorical laboratories also continued, however. The physiologist Michael Foster (1836–1907), for instance, in 1885 referred to "the chemical labour wrought in the many cellular laboratories of glands and membranes" (Foster 1885, 9); and in 1895 we find the cytologist and comparative anatomist Oscar Hertwig (1849–1922) explaining that "each living cell . . . resembles a small chemical laboratory [*einem kleinen, chemischen Laboratorium*], for the most varying chemical processes are almost continually taking place in it, by means of which substances of complex molecular structure are on the one hand being formed, and on the other are being broken down again" (Hertwig 1895, 126).

The metaphor of the cell laboratory or factory became particularly popular in physiological discussions of metabolism. The term "metabolism" was introduced by Schwann in his 1839 paper on the cell theory to denote the chemical changes carried out by the cells of living bodies (Schwann 1839, 193). The term derives from the Greek "το μεταβολικον," meaning that which is subject to or capable of initiating change (Liddel and Scott 1989). If one is interested in the physiological activities of the cell, how it builds up and breaks down materials from and into inorganic ones, then the comparison to a laboratory or factory is much more suggestive than the simple comparison to an elementary organism. For now the key question is not "From what ancient organisms did modern cells evolve?" but the more immediate ones, "How do these little things work?," "How do they manage to take in raw materials and reconfigure them so as to manufacture new sorts of products?" Answering these questions required the adoption of experimental investigations of cell activity guided by the materialist principles of chemistry and physics.

A turning point for the chemical investigation of cellular activity came in 1897 with Eduard Buchner's (1860–1917) demonstration of cell-free fermentation in a test tube. This was significant because Pasteur's work on fermentation had suggested the process was dependent on the presence of live organisms, yeast in particular. Buchner ground up yeast cells with mortar and pestle and added sugar to obtain a "press juice" in which fermentation took place. Buchner ascribed the cell-free catalytic activity to a soluble enzyme he dubbed "zymase" (Bechtel 2006, 97–98). For this and his subsequent research into the role of lactic acid in fermentation, Buchner received the Nobel Prize in chemistry in 1907. Because lactic acid, once thought to be a constituent of the alcohol fermentation process, was also detected in animal muscle after contraction, the chemical investigation of enzymes in cell-free conditions received a great stimulus.

But why suppose that employing the highly artificial laboratory techniques of physiological chemistry or biochemistry, *in vitro*, would be informative of events and processes naturally taking place in the whole cell and whole organism, *in vivo*? A simple response is that physiologists had few options but to learn as much as they could from their artificial investigations of the chemical activities taking place in living cells. However, if living cells were not entirely unlike human laboratories, if cells used the same materials in accordance with the same chemical principles, why should scientists working in their laboratories not be able to get a reasonably close understanding of what was going on in these miniature cell laboratories? The cell laboratory metaphor, there-fore, expressed both a hope that experimental investigation of cell-free chemical activity would prove fruitful, in addition to offering a kind of rhetorical justification for the approach itself.

Kohler (1982, 287) mentions that the physiological chemist Franz Hofmeister (1850–1922) depicted the cell as a biological machine shop or factory (*geordneten Betriebe*), with enzymes arranged on colloidal structures like machine tools (*Handwerkszeug; Zahnrad; Räderwerk*) on an assembly line (Hofmeister 1901, 19, 28, 29). The following year Hofmeister coined the term "biochemistry" with the creation of the journal *Beiträge zur chemischen Physiologie und Pathologie: Zeitschrift für die gesammte Biochemie (Contributions to Chemical Physiology and Pathology: Journal for Collective Biochemistry)*. The biochem-ist Rudolph Peters (1889–1982) mentioned hearing that Hofmeister "viewed the cell as a kind of complex micro-kitchen" (Peters 1937, 36–37), and wrote, "If we look with our eyes open we shall probably find several happenings which point to the presence of organized structure among enzymes in cells" (39). Extending the analogy between the cell and a laboratory or factory or kitchen, even, suggested that, not only does a living cell follow the same principles of chemistry as do chemists, but that it may also arrange its internal operations in an orderly spatial arrangement, as the chemist does her materials and equipment on the lab bench or as a chef arranges his ingredients and cooking utensils in his kitchen workspace.

This suggestion that the chemical activity of the cell was spatially as well as temporally organized was also promoted by Sir Frederick Gowland Hopkins (1861–1947). In his 1913 speech "The dynamic side of biochemistry" Hopkins described the cell as an "organized labora-tory" (Hopkins 1913, 715) in which the "chemical manufacture" of the "products" of metabolism occurs in stages, by the activity of small mol-ecules (an "army of enzymes" described as "agents," moreover) com-partmentalized in differentiated regions of the cell. This was in contrast to the prevalent assumption that metabolism was carried out in its en-tirety ("in block") by a complex macromolecule going by the name of "bioplasm." Kohler (1982, 63) describes the influence of this "machine shop" vision of the cell on younger researchers such as Sir Rudolph Pe-ters, then the Whitley Chair of Biochemistry at Oxford, who followed Hopkins's idea that metabolism must be ordered by cellular microstruc-ture (Kohler 1982, 91), and Martin O. Forster (1872–1945), whose President's Address to the Chemical Section of the 1921 Edinburgh meeting of the British Association for the Advancement of Science was titled "The laboratory of the Living Organism," in which the cell was described as a kind of laboratory.[7] Similarly, when J. B. S. Haldane (1892–1964) spoke in 1937 of the several separate genes responsible for the enzymes required to complete the synthesis of chlorophyll in maize, he described them as "controlling successive stages in the synthesis, much as a team of well-drilled PhD. candidates would do in a certain type of German laboratory" (Haldane 1937, 6).

Thinking of the cell as a laboratory or factory suggested that the cell interior ought to be a highly organized place, with something akin to an assembly line structure, a series of work stations at which various stages of the final products (proteins) were assembled. This was a very different conception from those who thought of the cell as a "bag of chemicals."[8] This presupposed that the chemical activity of enzymes occurred in a solution environment with little spatial organization and that interactions between ligands and the substrates on which they acted are governed by the laws of random molecular diffusion. But increasingly, scientists were beginning to believe that structure and order were necessary to explain the efficiency with which cells carried out their chemical work. As the Cambridge biochemist David E. Green explained: "In the cell we have evidence to believe that randomness is minimized–the organization of systems being such that each enzyme and component works at maximum efficiency. There is little cause for surprise therefore that *in vitro* reconstructions proceed at a very much slower velocity than the same components are capable of under physiological conditions" (Green 1937, 184). This shift in thinking about the internal organization of the cell's chemical activity is reflected in the third edition of Baldwin's *Introduction to Comparative Biochemistry*, with its added remark about the "modern concept of the cell as a highly organised . . . system" (Baldwin 1948, 141).

By the 1950s, the physiologist Leonard E. Bayliss (1900–64) (son of physiologist Sir William Maddock Bayliss), while commenting on cellular oxidation, noted the importance of subcellular "structure" with a telling mechanical analogy.

There is something in the cell structure, then, with which, or by whose aid, the work of the cell is carried on. There are arrangements by which the chemical energy of the oxidation processes is caught, as it were, before it has fallen to the state of heat. If we look upon the cell constituents as chemical compounds, merely without the assistance of some organised mechanism, nothing but heat could be obtained on oxidation. In the same way, if a petrol motor is smashed up and mixed together, with its fuel, nothing but heat would be obtained by burning the mass (Bayliss 1959, vol. 1, 354).

In time some investigators, like Boris Ephrussi (1901–79), were beginning to express frustration with biochemists who, trained in the tradition of the theory of solutions, treated cells as "bags of enzymes" (Ephrussi 1953, 108).[9] Bacteria, especially, were viewed this way because they lacked any internal structure that could be discerned with the microscope. It wasn't so much that biochemists were explicitly saying that cells are just bags of chemicals, but that their methods of investigation, which were reliant on cell-free homogenates, cell-presses, and cell-juices in test tubes, etc., were destructive of the living cell and its natural structure and internal organization, and so treated it as though it were equivalent to a bag of enzymes, even if they recognized this to be a simplification.[10] The rise of the new molecular biology throughout the 1940s to 1970s intensified the use of invasive and destructive techniques such as cell fractionation and centrifugation (Bechtel 2006). Living tissues were diced and chopped, the cells ground under great force and pressure to break open the plasma membrane, the contents suspended in solutions very different from their natural environments and spun in extremely high-speed centrifuges to sort the various particles and organelles into separate fractions by weight, so that specific chemical activity might be associated with the different cell components. And although these techniques were providing insight into the inner workings of cells, biochemists such as Efraim Racker (1913–91) asked, "Is our method of fractionation like the clumsy undertaking of a car mechanic who attempts to use his crude tools to analyze a watch? I believe that it is almost as bad as that. Nevertheless, we have no alternative and must hope that our tools will become refined as we proceed in the analysis" (Racker 1965, 89).[11] Leonard Bayliss also noted the doubts expressed by some biologists about the value of experimental analyses of metabolic processes carried out *in vitro* on "tissue slices, minced tissues and extracts from tissues" (Bayliss 1968, vol. 2, 172). Of course, *in vivo* studies would be preferable, Bayliss conceded, but if one insists that studies be carried out on the organism as a whole, he explained, one is restricted to observing inputs (ingesta) and outputs (excreta) only. "The comparison of the organism to a town where various occupations are carried on, is often made," he said; however, "If we notice that a large quantity of milk goes into the town and that a corresponding amount of cheese comes out we conclude that the milk has been used to make the cheese, but we learn nothing about the method employed" (172).

The problem then was how to employ methods that seemed to treat the cell as though it were little more than a repository for enzymes (a literal *cell* or *vesicle*) while at the same time maintaining that internal structural and spatial organization is important. The use of electron microscopy by researchers like Albert Claude (1899–1983), Keith Porter (1912–97), and George Palade (1912–2008) proved crucial for what Bechtel (2006) calls the *alignment* of biochemical function with cytological structure, by helping scientists to verify the identity of biochemically active components isolated in cell fractionations with structures found in intact cells. It also helped to reveal much more subcellular structure and organization than was possible with light microscopy alone.

The evidence obtained by electron microscopy confirmed the intuition of many scientists that the cell was a highly organized space, with important implications for its biochemical function. Peter Mitchell (1920–92) (Nobel laureate in Chemistry 1978) wrote in 1991 that "fifty years ago the bag-of-enzymes view of cell metabolism was prevalent and the chemical actions of metabolism were generally looked upon purely as processes of primary chemical transformation" (Mitchell 1991, 307). The biochemist and historian Joseph Fruton (1999, 322), however, does not credit this as an accurate or unbiased assessment of the enzyme group at the Cambridge Biochemical Laboratory where Mitchell obtained his PhD in 1951. Fruton (160) suggests the charge that biochemists regarded cells as bags of enzymes to be part of a mythology created by the community of physicists who turned to genetics after the Second World War, and that it was chiefly these geneticists who took up the "information theory of life," not biochemists, who regarded cells in this overly simplistic way (418).

The factory metaphor was also applied to the internal components of the cell by people like Albert Claude, one of the pioneers of the molecular biology revolution, who, according to the recollection of Rollin Hotchkiss (1911–2004), was referring to mitochondria as the "factories of the cells" as early as 1942 (Moberg 2012, 76). In his 1948 Harvey lecture, Claude introduced what has since become a standard and almost clichéd idea when he pronounced "mitochondria may possibly be considered as the real *power plants* of the cell" (Claude 1948, 137; italics added). This set in motion the modern understanding of molecular cell biology, wherein subcellular organelles are recognized as the localized sites of biochemical activity, much as the synthesis of commercial products is organized in an industrial factory. Philip Siekevitz

(1918–2009), a biochemist who worked in the Porter-Palade lab at the Rockefeller Institute, summarized the new conception of subcellular architecture and organization this way:

> We have come a long way from the time when a cell was considered a bag of loose substances freely interacting with one another. The cell, like the mitochondrion, has a rigorous compartmented organization. Perhaps this is not surprising. When we build a factory, we don't park its raw materials and machines at random. We arrange matters so that the raw materials are brought in near the appropriate machines, and the product of each machine is efficiently passed along to the next. Nature has surely done the same in the living cell (Siekevitz 1957, 144).

The emphasis on the importance of spatial organization and structure in the inner cell environment saw perhaps its most striking achievement with Peter Mitchell's chemiosmotic hypothesis of adenosine-triphosphate (ATP) synthesis. The process whereby the mitochondria serve as the cell's "power plant" is called oxidative phosphorylation, which involves the oxidation of carbohydrates to release the energy stored in the molecular bonds of the organic material consumed (food), and using that to create ATP, a more readily available form of energy "currency." The standard view throughout the 1940s and 1950s was that a soluble high-energy intermediate existed between oxidation and the final step of phosphorylation (whereby a phosphate is transferred to a molecule of adenosine-diphosphate (ADP) to create the final product ATP). But when attempts to find the soluble intermediate were unsuccessful, Mitchell proposed in 1961 that it did not exist, and that the energy required for the phosphorylation of ADP to ATP was obtained rather by the buildup of a gradient of electrically charged protons across the inner membrane (the cristae) of the mitochondria. After much controversy, the hypothesis was confirmed and Mitchell received the Nobel Prize for chemistry in 1978. Grote (2010, 187) argues that the chemiosmotic hypothesis represented a novel research style in bioenergetics that clashed with the established chemical theory by suggesting that cellular reactions occur in association with spatial structures such as membranes and other surfaces rather than in a simple aqueous solution. Metaphors and analogies of spatially organized systems such as factories and batteries played a significant role.

The factory metaphor was also applied to investigation of protein synthesis in the early 1960s. Alexander Rich (1924–), a biophysicist at

MIT, and his team discovered in 1963 polyribosomes or polysomes—clusters of ribosomes which simultaneously "translate" a strand of messenger RNA into protein. In a 1963 *Scientific American* article, Rich wrote, "Within the past 18 months experiments in our laboratory at the Massachusetts Institute of Technology and elsewhere have led to the hypothesis that the protein 'factories' of the cell are not single ribosomes working in isolation but collections of ribosomes working together in orderly fashion as if they were machines on an assembly line" (Rich 1963, 44). This picture of the cell as composed of tiny molecular machines geared to the manufacture of various products has become standard to molecular biology.

The revolution in molecular biology was of course inspired by developments in molecular genetics, guided by the metaphors of DNA as the "code" for protein synthesis and the "blueprint" for the development of new organisms (Kay 2000; Keller 1995; 2000; 2002). The creation of molecular-biochemical techniques such as recombinant DNA and gene-splicing made it possible to manipulate and to improve the efficiency of the cell's natural operations for specifically human purposes. Now, in the modern era of biotechnology, the metaphor of the cell factory has been literalized. In 1999 the European Union launched a 400-million-euro research program called "the Cell Factory Key Action."[12] A supporting document explains the guiding vision for the project.

> The concept of the "bio-product" is as old as the knowledge involved in the making of bread, beer, wine or cheese. However, recent techniques and knowledge in molecular biology and genetics mean that living cells — from bacteria to man — are now becoming real "factories." In vast fermentation vats, engineers can direct and control natural metabolism in order to produce all sorts of substances with a high added value: proteins, amino acids, alcohols, citric acid, solvents and even bio-plastics. This industrial mastery of the mechanisms of life opens up revolutionary perspectives in the development of new kinds of medicines, foodstuffs with specific nutritional properties, and bio-degradable biochemical products (Aguilar 1999, 121–24).

Today there are professional journals specifically devoted to research into cell factories. A recent article published in one such explains further the opportunities for the engineered improvement of the cell's innate manufacturing ability.

> Genome programs changed our view of bacteria as cell facto-
> ries, by making them amenable to systematic rational improve-
> ment. As a first step, isolated genes . . . or small gene clusters are
> improved and expressed in a variety of hosts. New techniques
> derived from functional genomics . . . now allow users to shift
> from this single-gene approach to a more integrated view of the
> cell, where it is more and more considered as a factory. One
> can expect in the near future that bacteria will be entirely re-
> programmed, and perhaps even created de novo from bits and
> pieces, to constitute man-made cell factories. This will require
> exploration of the landscape made of neighbourhoods of all the
> genes in the cell. Present work is already paving the way for that
> futuristic view of bacteria in industry (Danchin 2004, 13).

The idea that the cell is a chemical factory has become somewhat a
dead metaphor, but a very lively and potentially lucrative economic enterprise, as biotech engineers learn how to "rationally improve" cells
by "reprogramming" them to produce "high value-added carotenoids,"
"proVITamin A," and other "cell factory crops."[13] Wöhler's and Liebig's
early insights into the chemical basis of cell synthesis are now in massive
industrial application.

Articles such as Cabeen and Jacobs-Wagner (2010) —"A Metabolic
Assembly Line in Bacteria"— attest to the continued relevance of the
factory metaphor for thinking about the spatial organization of what
is commonly called the "enzyme equipment" of cells.[14] That biologists
look to the concept of a factory as more than a rhetorical device for
communicating with nonscientists is evident from articles such as Giuseppin, Verrips, and Van Riel (1999): "The cell factory needs a model of
a factory."[15] It seems reasonable to argue on the basis of these developments that metaphors, far from being merely useful heuristic devices or
convenient modes of conveying difficult scientific concepts to a popular
audience, can act as a kind of conceptual instrument allowing scientists to think about their subject of study in novel and fruitful ways. As
conceptual instruments, metaphors function as analytic tools that help
scientists take apart (conceptually) a system –like a cell—and suggest
further avenues of investigation.[16] If cellular products (e.g., proteins)
are being assembled in stages inside the cell as they are in a factory,
then one is naturally prompted to wonder whether the product is carried about from one stage of assembly to another, and if so how? Is
there something like an assembly line or conveyor belt for carrying the

subcomponents of the finished product from one location to the next? This has helped cell biologists to rethink the initial idea that the cell is a fluid bag of loosely floating chemicals to something more like a rigidly organized factory floor, where the tools and workbenches for assembly of related parts must be kept relatively close to one another, hence the thinking about the function of the cytoskeleton and various dynein motor tracks, etc., for carrying molecules about the cell to their required places.[17]

But beyond their immediate and obvious power to reorganize how we think about things in the real world, a metaphor may also function as a kind of "experimental technology" capable of effecting real material change on its object, as the history of the cell factory metaphor attests. For it is with application of the factory metaphor that cells have been materially altered. It is possible, of course, that bacteria and other cells would have eventually been manipulated by bioengineers to become more efficient producers of various bioproducts without the factory metaphor, but it is also clear that the metaphor expedited this transition by making it a natural and commercially plausible project to pursue.

The cell factory metaphor has become so familiar today that it may appear almost inevitable. In fact, it has become an integral element of science pedagogy in North America through an initiative of the American Association for the Advancement of Science (AAAS). School children in grades 6–8 are asked to think of ways that the cell is like a factory, and in variations of this project they are encouraged to create their own original analogies and metaphors for the cell.[18] The chemist Theodore Brown recounts a colleague's confusion when confronted with the widespread use of the factory metaphor for talking about the cell: "How else could you talk about it?" was the puzzled response (Brown 2003, n. 1, 207–8). To make the point that the factory metaphor is not inevitable, Brown asked his colleague to consider how biologists unfamiliar with the existence of factories (contingent social and economic institutions that they are) might attempt to understand the cell. Why, for instance, aren't people talking about the cell as a kibbutz?[19] The question helps emphasize that the contingent social experiences of the people doing science in any given age and nation can influence the metaphors and analogies they will come to adopt and find useful. It is no doubt relevant that the scientists discussed in this chapter were all members of industrialized nations during a period of rapid economic growth and activity, i.e., Europe and North America during the nineteenth and twentieth centuries. Indeed, the cell factory metaphor has an industrial-commercial connotation that is lacking in the cell laboratory metaphor, since a laboratory can still refer to a location where a scientist might be occupied with the disinterested pursuit of truth rather than hopes of patent applications and commercial wealth. Though Wöhler was working in an academic lab, Liebig turned organic chemistry into a commercially lucrative applied science with the creation of the synthetic dye industry. In our current political and economic environment, where scientists working at cash-strapped publicly funded universities are increasingly encouraged to forge partnerships with private for-profit businesses and the commercialization of research is euphemistically touted as "knowledge-transfer," talk of cell factories seems quite natural and fitting. The broader implications of this industrialization of living cells on a social and even ethical level receive little discussion.

Not only have cells become more factory-like, but in illustration of what Max Black called the "interaction" theory of metaphor, factories are also becoming more cell-like. Black (1979) urged that when two things are linked through a metaphor, e.g., "Man is a wolf," the conceptual traffic is not solely one-way, for an interaction takes place between the two subjects so that not only does the target (man) come to seem like the source (wolf) but wolves also come to be regarded as having more human qualities. And indeed just such an interaction has occurred in the thinking of those who plan and organize industrial factories. The theory of "cellular manufacturing," with its notion of "work cells" and "cellular assembly," uses the arrangement and internal organization of biological cells as a model for the layout and organization of materials and workers for the industrial manufacturing of various products. "Cellular manufacturing, sometimes called cellular or cell production, arranges factory floor labor into semi-autonomous and multi-skilled teams, or work cells, who manufacture complete products or complex components."[20] Who can say where the influence of a metaphor will end?

3. The cell as a machine

Laboratories, factories, and motors are all examples of human technology, but one of the most familiar and most general forms of technology is a machine. This likely explains why the concept of a machine has been such a pervasive background metaphor in cell and organismal biology. The practice of construing living organisms as machine-like is an offshoot of the mechanical philosophy of the sixteenth and seventeenth centuries. One of its most significant proponents was René Descartes (1596–1650), whose *Treatise on Man* (1662) models the human body and its physiology as a complicated machine consisting of hydraulic pipes, tubes, valves, cords, and other contrivances. Much has been written about the mechanical philosophy and its applications in biology.[21] The metaphors and analogies favored at any given historical period tend to reflect the technological innovations of the day, and since Descartes's time these have varied from heat engines to the electronic devices of our current digital computer age. My discussion of machine metaphors will for the moment be restricted to the period ranging from the nineteenth century to the first half of the twentieth, just prior to the dawn of the computer age. Computer and electronic engineering metaphors will be the focus of the discussion of cell signaling theory in chapter 4.

A key motivation for thinking of living organisms and cells in machine terms has been the assumption that, whatever other principles may prove necessary to explain life, living things ought to be subject to the same physical and chemical laws as inanimate matter. In the nineteenth century, this meant situating living things within the framework of thermodynamic laws: specifically, the law of the conservation of energy and the law of the dissipation of energy or entropy increase. Living cells, consequently, were assumed to function like any human-made machine (e.g., a steam engine) in that they must draw upon an energy source for their metabolic and reproductive activities and use this energy to work against the organization-destroying trend of entropy increase. In opposition to this is the rival position of vitalism, which assumes that living organisms and cells have access to some nonmaterial force or principle and, as a consequence, cannot be understood solely in material physical-chemical terms.

Although materialism had significant support in the nineteenth century, not all of its advocates were entirely in agreement about the adequacy of mechanical or machine analogies for explaining how living organisms and cells operate. For this reason, Garland Allen (2005) has proposed that a distinction be made between two forms of mechanism or mechanistic thinking. What he calls "philosophical mechanism" is the materialist philosophy that construes living organisms as merely complicated machines; while "operative or explanatory mechanism" is an epistemic approach to the explanation of a living system that seeks to identify the salient parts and their causal effects on one another, without necessarily assuming that the causal relations involved will be precisely like any sort of human-made machine. Philosophical mechanism carries, then, a greater ontological commitment to the claim that organisms (and cells) are really nothing more than very complicated machines composed of material parts in causal interaction with one another; whereas explanatory mechanism need be only a heuristic or methodological position that leaves open the possibility that living systems are in reality much more complicated than any machine, but takes the analytic-experimental approach as the best hope for untangling the complex multilevel interactions operating in living systems. It is in this broader and looser sense, for instance, that biologists may speak of natural selection as a "mechanism" or search for a causal mechanism underlying angiogenesis in the development of a cancerous tumor.

A good illustration of the operative or explanatory form of mechanism is provided by Rudolf Virchow's 1858 lecture "On the Mechanistic Interpretation of Life." Here Virchow insisted that a unified approach in scientific investigation demands the premise that there are universal laws governing the regular activity of cause and effect in biology, and that no invocation of vital forces inconsistent with established physical and chemical science are to be admitted. The cell is to be recognized as the essential site of all life activity and is to be understood as a material system no different in its essential physical-chemical features than any nonliving inorganic matter, except that through their peculiar arrangement and confinement within the minute enclosed space of the cell membrane, these physical and chemical forces result in the peculiar properties of life. "The same kind of electrical process takes place in the nerve as in the telegraph line or the storm cloud; the living body generates its warmth through combustion just as warmth is generated in the oven; starch is transformed into sugar in the plant and animal just as it is in a factory" (Virchow 1958[1858], 107). As much as this sounds like a straightforward espousal of materialism, Virchow rejected such a conclusion. "*In fact, however, the mechanistic interpretation of life is not materialism*" (108, italics in original). Materialism, he says, is a speculative system of philosophy that presumes to go beyond the necessarily fragmentary evidence of experience, and Virchow seems to align himself with an empiricist or antimetaphysical positivism that would disallow any claim to knowledge about the fundamental essence of things. Virchow's mechanistic conception of life seems to come down to the statement "*The law of causality applies to organic nature as well* [as to inorganic nature]" (108, italics in original), meaning that a scientific approach assumes no biological phenomenon to be a miraculous event that is in principle inexplicable from within a framework of regular causal influences involving purely natural forces. Virchow did, however, on occasion describe his position as a form of vitalism, most notably in the 1856 essay "Old and New Vitalism." But as Lelland Rather has explained, Virchow's vitalism did not involve the supposition of any

nonphysical forces or guiding entelechy, as was favored by the "old vitalism" of the Montpellier School or the German *Natur-philosophen*
(Rather 1990, 29, 78–80).[22] What Virchow intended by calling his position vitalism was that the processes peculiar to life (and so to living
cells) cannot be reduced to physical and chemical principles alone, but
must be attributed to their arrangement and organization within the
living cell, and though Virchow considered it very likely that life originated at some time in the distant past from nonliving matter, evidence
suggested that spontaneous generation no longer occurs, so that presently all living cells arise from previous living cells (*omnis cellula a cellula*). Life, in other words, comprised a set of properties and capacities
restricted to material systems having the organization of a cell.

On this point Virchow is to be contrasted with his colleagues, the
Berlin "Medical Materialists": Emil Du Bois-Reymond (1818–96),
Hermann von Helmholtz (1821–94), Ernst von Brücke (1819–92), and
Carl Ludwig (1816–95), who like him were students or associates of the
influential physiologist, comparative anatomist, and vitalist Johannes
Müller (1801–58). This quartet of physiological reductionists vowed to
banish biological science of all vitalist mysticism and endorsed a program wherein physiology would be reduced to the attractive and repul-
sive forces of material molecules and atoms (Coleman 1977, 151*ff.*).
Brücke's thinking about cells ran very close to Virchow's, for though
he never described himself as a vitalist, he too believed that so far as
the evidence suggested, life was a property peculiar to the cell level of
organization, not to be found at any lower level.

It is to capture such subtle distinctions within the materialist camp
that Allen has proposed the labels mechanistic and holistic materialism
(Allen 1978; 2005). *Mechanistic materialists* approach the organism as
an aggregate of separate parts that are best understood when studied
in isolation from one another under controlled circumstances. A living
organism is to be studied as one would any machine: by opening it up,
identifying the separate parts, removing them for closer inspection, and
then seeing how they causally interact with one another to create the
behavior of the whole. There is an assumption here that the organism is
the sum of its parts. *Holistic materialism*, on the other hand, may also
involve analysis of a living system into smaller parts, but on the holist
view, the parts themselves can be properly understood only in their relations to one another and to the system as a whole. This is because
living organisms, unlike machines, are characterized by emergent properties arising from the interaction of parts and different levels of organization, and these emergent properties can neither be predicted nor

explained from the investigation of isolated parts on their own (Allen 2005). This means one cannot expect to identify the natural properties of the complete system or organism in its normal state by investigation of its components in isolation from one another, through experimental manipulation of one small component at a time to tease out its inherent properties; for according to this view, the whole is more than the sum of its parts. Holistic materialism (also known as *organicism*) emphasizes the importance of the organization of organismal components and how their interactions can have transformative effects on the parts not seen in machines, whose parts remain essentially unchanged by their causal interactions with one another, aside from the destructive effects of wear and tear.

Consequently, one needs be careful not to assume that because a scientific figure uses the language of machines when talking of organisms that they are *ipso facto* philosophical mechanists or mechanistic materialists. For instance, the physiologist Claude Bernard, who was criticized by some of his peers for advocating a form of vitalism, also used mechanistic language: "Now, a living organism is nothing but a wonderful machine endowed with the most marvelous properties and set going by means of the most complex and delicate mechanism" (Bernard 1961[1865], 91). Yet, like Virchow and Brücke, Bernard insisted on the importance of the mutual interactions of the cell, tissue, and organ units on one another in their creation of a *milieu intérieur* effectively staking out an irreducibly biological approach to the study of physiology that employed the principles and methods of chemistry without supposing that the phenomena of interest are wholly reducible to chemistry and physics. "For physiologists, the truly active elements are what we call anatomical or histological units . . . [and] there can be no question that these histological units, in the condition of cells and fibres, are still complex. That is why certain naturalists refuse to give them the names of elements and propose to call them elementary organisms" (101). Holistic materialists like Virchow, Bernard, and Brücke therefore challenged the metaphor that the cell is the biological equivalent of the chemical or physical atom or element insofar as that suggested a condition of simplicity of structure or lack of internal organization or complexity. It is for this reason that late-nineteenth-century advocates of the cell theory like Oscar Hertwig described the cell as a "marvellously complicated organism, a small universe, into the construction of which we can only laboriously penetrate by means of microscopical, chemico-physical and experimental methods of inquiry" (Hertwig 1895, ix). Cells, therefore, may be the elements or building blocks of higher plants and animals, but that was not to say they were themselves simple; nor was it implied that they are reducible to the molecular forces of their protoplasmic substance, not if cells are in fact the true elementary units of life, for that would be to ascribe life to the subcellular molecular components.[23]

The clearest example of philosophical mechanism and mechanistic materialism in the early twentieth century is Jacques Loeb (1859–1924), the physiologist and champion of the "engineering ideal" in biology (Pauly 1987). In his lecture "The Mechanistic Conception of Life," delivered in 1911 at the First International Congress of Monists in Hamburg, Loeb laid out his program for reducing all vital phenomena (growth, metabolism, reproduction, heredity, movement, irritability, and behavior) to physicochemical principles in such a way that the biologist would be able to experimentally manipulate and control living systems as an engineer designs and controls machines and other technological devices (Loeb 1912). Fertilization and reproduction were already well on the way to being explained on physicochemical principles, he insists; and even what is taken to be the "inner" life of psychic phenomena, he argued were in principle explicable along the lines of plant and animal tropisms (the inevitable response of organisms and cells to natural forces such as light, gravity, and electricity). The scientist, he urged, is an engineer whose job it is to control nature, and so biology must be less descriptive like natural history and become more rigorous, more experimental and quantitative, like the physical sciences. In the essay "Mechanistic Science and Metaphysical Romance" (1915), Loeb urged biologists to seek out causal mechanisms that can be manipulated at will to produce anticipated effects, and proposed that by making precise quantitative measurements and predictions science will eventually settle down to the correct mechanistic picture of the atomic and molecular reality underlying the facts of biology. In *The Organism as a Whole: From a Physicochemical Viewpoint* (1916) Loeb further articulated his argument against vitalists like Bernard, Driesch, and Jacob von Uexküll (1866–1944) by undermining their claim that a fertilized egg required some kind of guiding force or entelechy, or "supergene" to arrange itself into a harmonious and purposeful whole. This might be the case, Loeb conceded, if the egg was a structureless, homogenous blob of cytoplasm, but in fact, he explained, experimental evidence on artificial parthenogenesis—by which he showed how manipulating the chemical properties of the seawater in which the eggs of sea urchins and other marine invertebrates were placed, they could be spurred on to nearly normal development—strongly suggested that the unfertilized egg itself was a future embryo reacting to regular physicochemical principles. No male gamete was even necessary in these cases for the development of a new organism, let alone any nonmaterial or vital guiding force.

Following Loeb's lead, the Cambridge zoologist James Gray (1891–1975) set out, in his influential *Textbook of Experimental Cytology* (1931), to put cytology on firmer chemical-physical foundations. This, as we may now anticipate, involved the use of machine language. He wrote:

> In cell physiology we are, in fact, attempting to dissemble the machinery of an organism into its simplest component parts. When we know how each of these works in an isolated state we shall be ready to integrate the data and gain some conception of the whole organism . . . That we cannot thereby obtain an adequate picture of all vital activities will be only too obvious, but even a cursory summary of recent physiological work will show a growing need for a precise knowledge of intracellular processes. When the organism as a whole establishes with its environment an equilibrium of profound biological significance, it does so by the machinery of its individual cells (Gray 1931, 5).

Gray then elaborated on this mechanistic view of the cell with regard to the problem of the energetics of cell respiration (which was then still little understood) by writing, "We can say that in some respects the cell behaves as though it were a self-charging accumulator which is constantly maintaining a leaking condenser" (5). Gray then conceded that "analogies are dangerous concepts, but it looks as though the only real conception of a living cell as a dynamic unit is provided by comparison with suitable types of inanimate machines" (5). The cell, of course, is a peculiar sort of chemical machine that relies on both aqueous solution and macromolecular proteins such as enzymes. As a cytologist and zoologist, Gray had concerns about the limitations of a purely chemical approach to cell function. "The isolation of enzymes," he wrote, "is equivalent to the isolation of parts of the working machine, and sooner or later we may be in a position to put some of these parts together. We seem driven to the conclusion that a real understanding of cell structure will rest on a knowledge of how the various parts of the cells are orientated in respect to each other, rather than on a study of the chemical constitution of these parts" (30). Biochemists, he suggested, would likely be forced to pay greater attention to the spatial "orientation of protein molecules one to another" (31), and he then actually used the machine analogy to insist on the importance of taking a more holistic approach: "If we are engaged in dissembling an aeroplane into its component parts we cannot expect to rebuild them into an effective machine until every single constituent is known to us both in form and function"; however, the difficulty of understanding how the parts function harmoniously together at the level of a complete cell, let alone a whole organism, cannot be resolved by a wave of the "magic wand of chemistry," and, he warned, "A biological problem disguised by the sparkling terminology of the chemist is too often a pathetic and rather disreputable object" (31).

But biochemists were themselves expressing similar concerns about the adequacy of the mechanistic approach. The biochemist David E. Green, who did important work on the role of enzymes in oxidative phosphorylation, wrote in 1937:

> The mastering of a particular machine requires not only a knowledge of the component parts but also the practical ability to take the machine to pieces and reconstruct the original. Obviously a resynthesis of the component parts is impossible without an appreciation of their functional and spatial interrelationships. The biochemist has made great progress in the way of describing the chemical constituents of the living machine and the transformations which they undergo, but he has only begun to resolve the living unit and to imitate some of its activities by permutation and combination of the component materials *in vitro* (Green 1937, 175).

The biochemist can extract various parts of the cell's machinery, and with them synthesize functional compounds, Green explained, but one may doubt whether these creations have much resemblance to the natural state of the cell's inner workings.

> A sufficiently ingenious mechanic could separate the parts of a baby Austin and use them to make a perambulator or a pressure pump or a hair-dryer of sorts. If the mechanic was not particularly bright and was uninformed as to the source of these parts, he might be tempted into believing that they were in fact designed for the particular end he happened to have in view. The biochemist is presented with a similar problem in the course of his reconstructions . . . There is thus a grave element of risk in trying to reason too closely from reconstructed systems to the intact cell. The reconstruction can have no biological signifi-

cance until some definite counterpart of these events is observed *in vivo* (Green 1937, 185).

What is notable from remarks such as these is that the machine metaphor not only provided scientists with a suggestive model of how to approach the study of cell function, but that it also allowed them to understand its limitations in language and imagery that was clear and accessible. The analytic approach continued to be productive in the decades to follow, even as the preferred metaphors changed during the 1950s and 1960s from mechanical machines to ones drawn from electronic engineering and computer systems. As scientists made progress in opening up the black box of the cell nucleus with the techniques of molecular genetics, they began to speak more of the genetic code and of DNA "instructions" and blueprints.

Despite the prevalence of high-tech computer metaphor in modern biology, there has also been a strong trend within molecular biology to describe and think about proteins in particular as being very machine-like, as illustrated by Bruce Alberts's (1998) editorial piece "The Cell as a Collection of Protein Machines," which advocated training future generations of cell biologists to approach the cell as engineers do machines. Alberts opens his opinion piece by commenting on how biologists no longer view the cell as a fluid compartment in which enzymes fortuitously meet their respective substrate-partners by random diffusion (i.e., a bag of enzymes), but as highly ordered spaces in which work is compartmentalized according to various stages of activity (as it is in a factory).

Allen (1978, 183–84) states that the history of biochemistry (like other aspects of biology) reveals the importance of a stage of mechanistic thinking, during which time enzymes were considered analogous to machine parts, functioning rather independently of their particular environmental surroundings; but that once scientists had sorted out the "cast of characters," as it were, a stage of holistic materialism was required to understand more adequately how these parts respond to changes in their environment that they themselves help to create. (This theme will be central to the next chapter.) Even today, articulation of the shortcomings of a mechanistic approach continues to make use of homey analogies like the ones mentioned above by Gray and Green, as seen in the rather humorous article by Lazebnik (2002), "Can a Biologist Fix a Radio?– Or, What I Learned While Studying Apoptosis." In contrast to the unmethodical experimental approach employed by cell biologists, which Lazebnik likens to yanking out or silencing various components of a radio, whereby biologists "silence" or "knock out" various cell components to discover what is for a time considered to be the most important causal component of an entire system (represented by diagrams placing this one component at the center), what is needed, Lazebnik argues, is a rigorous and standardized language of the variety achieved by engineers, who can peer into the back of any transistor radio and effectively communicate with one another the variety and nature of the components involved because of the agreement on terms and conventions for diagrammatic representation of systems of interest. Cell biology, according to this critique, actually needs to become more like electrical engineering.

To this end, a group of synthetic biologists at MIT founded in 2006 the "BioBricks" project, with the object of creating standardized DNA sequences that can be mixed and matched with predictable results allowing scientists to "reprogram" cells and organisms to achieve various bioengineering projects such as the synthesis of biofuels and medicines.[24] This is the next step in the rational engineering of cell factories. In 2012, a spin-off project from this same group was created, known as the International Genetically Engineered Machine (iGEM) Foundation, whose mission is to promote education about and advancement of synthetic biology.[25] The official websites of these projects display visual metaphors on the cell-machine trope. The primary intention of such visual metaphors is undoubtedly rhetorical, to help visitors to the websites understand the basic approach and objectives of the synthetic biology project. But the cognitive power of these metaphors should not be underestimated either, for they provide in a very simple and effective way a guiding vision of what it is the scientists hope to achieve. In a few words (or images), the metaphors (verbal and visual) express the motivation for the entire project and so serve to guide the expenditure of time, money, resources, and cognitive effort. While perhaps not exactly constitutive of the project, in the sense that the science could be done without use of the machine metaphors, the science is only imperfectly understood without attention to the metaphors that guide it.

Garland Allen explains this sort of approach works because: "While organisms are not machines, within narrow limits they can be forced to function like machines" (Allen 2005, 280). But it is in the interest of understanding those features of the cell or organism lying outside of those narrow limits that other biologists have insisted on the necessity of a more holistic or organicist approach. The trouble associated with any metaphor is that it has the potential to become self-legitimating. As Richard Lewontin said of the machine metaphor: "We cease to see the world *as if* it were *like* a machine and take it to *be* a machine. The result is that the properties we ascribe to our object of interest and the questions we ask about it reinforce the original metaphorical image and we miss the aspects of the system that do not fit the metaphorical approximation" (Lewontin 2000, 4). Perhaps even more worrying than the possibility that the metaphor may prevent us from recognizing those aspects of the cell that are nonmachine-like, is the reality that use of the metaphor turns the cell into a literal machine or factory. Metaphors have the power, in other words, not just to bring about an *epistemic* shift in how we *think about* cells, but an *ontological* shift in the *very nature* of cells. Once cells have been reprogrammed, re-wired or otherwise reengineered, they may behave in ways more artifactual than natural. So while achieving the ability to take cells apart and put them back together in various functional ways may be a sign of progress in understanding how cells *can* work, like the Austin Mini and other engines in the analogies of the biochemists discussed above, this may not signal an understanding of how they *do* work in their natural states—for the natural states of cells, unlike machines, are highly variable, dynamic, and responsive to changes in their environment. It also goes without saying that the legal, social, and philosophical implications of these bioengineering projects remain to be seen.[26]

There is yet another reason why some philosophers and scientists have criticized the use of machine metaphors in biology, and this has more to do with politics and science education. Massimo Pigliucci and Maarten Boudry have expressed concern that when biologists talk about cells and organisms using machine and engineering language, it only provides fodder for Intelligent Design advocates, who take the metaphors literally.[27] ID theorists argue that if the cell is a complicated machine or factory, consisting of elaborate signaling circuits and protein machines, then there must be an intelligent engineer responsible for their design and creation. In addition to this more political motive, Pigliucci and Boudry also suggest that the ruling master metaphors of machine and information analogies (genetic blueprint, genetic instructions, reverse-engineering organisms and cells, etc.) have misled scientific research and held up further progress.[28] They write:

> Wittgenstein . . . famously said that "Philosophy is a battle against the bewitchment of our intelligence by means of our language." Perhaps a contribution of philosophy of biology to the field of synthetic biology is to help free the scientists from the bewitching effects of misleading metaphors, so that they can

simply get on with the difficult and unpredictably creative work
lying ahead (Boudry & Pigliucci 2013, 668).

It is not clear whether Boudry and Pigliucci believe that scientists
need—or indeed can—break entirely free of metaphor or whether they
need only abandon the metaphors they believe to be misleading. But
their concern over the consequences of the use of metaphor in science
is, as we shall see below, shared by scientific investigators who are themselves working at the forefront of cell biology; which adds to the case
that philosophers of science should take the issue of metaphor more
seriously than they have been inclined to do.

Conclusion

Scientists interested in understanding how cells manage to function as
they do look for mechanisms, i.e., causal accounts of how the various
parts of the cell interact to produce the various effects and phenomena of the whole cell as a system. Metaphors of factories, engines, and
machines generally provide analogical guides to this analytic approach
by suggesting the type of parts to look for and the sort of interactive
behavior they may be engaged in, as well as the physical principles that
may describe their causal interactions. Of course, a living organism is
not really or exactly like a plane, a bike, an engine, or a radio, etc., but
how else could we proceed if we want to understand not just what living things do but how they do it? A perhaps unintended consequence
of thinking of cells "through the lens" of a particular metaphor is that
they will actually be reconstructed according to that image, as we saw
was the case with the idea that the cell is a chemical factory, and as we
will see further in chapter 4 with the practice of thinking of the cell as
an electronic device composed of circuits, switches, and other accoutre-
ment borrowed from computer engineering.

The position Allen calls mechanistic materialism assumes a dynamism of the whole system—there is a recognition that there is a constant dynamic interaction *between* the parts. It is this dynamic communication of causal energy from part to part, as is found in a machine like
a mechanical clock (with a coiled spring communicating kinetic energy
to a drive wheel, which by direct physical contact communicates this
motion to another cog wheel, etc.) that accounts for the cell's activity.
But holistic mechanism or organicism recognizes another level of dynamism, a dynamism *of the parts themselves* that is unlike anything seen in

machines. For cells and cellular components undergo dynamic changes or transformations as a result of their mutual interactions, which makes analogies with human social agents and how we are personally affected by our social interactions with one another more pertinent for understanding the internal causal mechanisms of the cell. It is for this reason that social metaphors have proven attractive in cell biology. This is the topic to which we turn next.

3

Cell Sociology: The cell as social agent

> No systematic attempt . . . has been made to apply to cellular societies the analytical approach which has been so successful in the study of animal social interaction: that is, to take the cell as a unit and to investigate how its behavior is influenced by other cells.
>
> **Abercrombie and Heaysman 1953, 111**

1. Introduction

We have noted that cells are commonly portrayed in two chief ways: as artifact or as agent. According to the first, a cell is described metaphorically as a space or chamber defined by a surrounding wall or membrane (this being the original meaning of the term "cell"), as the building stones or blocks (*Bausteine*) from which living organisms are constructed, or as a kind of machine or factory. According to the second, the cell is a living organism in its own right. It is an elementary-organism (*Elementarorganismus*). This, as we saw in chapter 1, made popular the idea that a plant, animal, or human represents a society of cells or "cell-state." Accordingly, cells are not only morphological and physiological units but social units too. The idea of the cell society continues to be popular today (see Sonnenschein and Soto 1999; Wolpert 2009; and fig. 3.1) and it exerts significant influence in framing theoretical discussions about the evolutionary origins of multicellu-

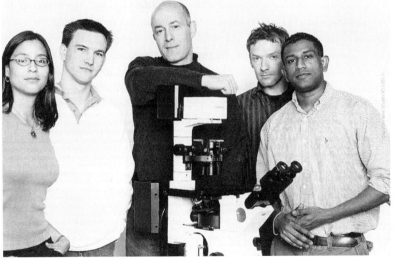

SEEKERS IN THE SOCIETY OF CELLS.

At Dr. Simon Watkins' lab, they look at cells the way anthropologists look at human culture: as communities of good guys and bad guys, of traders and communicators, of connections and relationships. "We are the observers," Simon says. "We never jump to conclusions. We let the conclusions jump to us." His mantra? "Imaging is everything." Which is why the best and the brightest of tomorrow's seekers and solvers find their way to Pittsburgh and the Watkins Lab.

OLYMPUS MICROSCOPES. **ROCKET SCIENCE™.**

Find out more about Olympus microscopes at www.olympusamerica.com/microscopes 800-455-8236

OLYMPUS

For Free Info, Circle 15

(From L to R)
Ana Bursick · Research Specialist
Stuart Shand · Research Specialist
Simon C. Watkins, Ph.D. · Director
Glenn Papworth · Research Associate
Romesh Draviam · Graduate Student
Center for Biologic Imaging,
University of Pittsburgh Medical School,
Pittsburgh, PA

FIGURE 3.1 The body as a society of cells. Olympus microscope ad. Text reads: "At Dr. Simon Watkins' lab, they look at cells the way anthropologists look at human culture: as communities of good guys and bad guys, of traders and communicators, of connections and relationships." (Reprinted with permission from Olympus Corporation of the Americas Scientific Solutions.)

larity (Reynolds 2017). But is it anything more than a catchy metaphor? Does it do any real cognitive work? This chapter outlines the history and underpinning philosophy of approaches in cell and organismal biology that treat the cell as a social organism and multicellular organisms as cell societies, an approach some biologists have described as "Cell

Sociology." This social conception of cells and organisms—especially in the area of developmental biology—constitutes an important means of appreciating the complex and modular nature of living things.

2. Alexis Carrel and the "new cytology": introducing "cell sociology"

Advocates of the cell standpoint celebrated a significant victory when in 1907 Ross G. Harrison reported his successful growth of nerve cells from segments of amphibian embryonic tissue. Harrison's experiments were noteworthy for finally resolving a longstanding dispute over whether the nervous system was composed of individual nerve cells (neurons) or a continuous network of protoplasmic fibers. But the technique he used to grow nerve cells in vitro proved to be of even greater significance. Harrison adapted a technique that had been used since the 1880s to grow bacteria in cultures (referred to as the "hanging drop" method) to observe with a microscope the growth and behavior of animal cells living outside the body. By the 1920s, improvements had been made in what were called tissue and cell culture techniques, led in particular by the surgeon Alexis Carrel, who famously claimed to have maintained a line of embryonic chicken heart cells alive in vitro for more than thirty years (Landecker 2007). Despite there being genuine grounds for skepticism about Carrel's own claims, tissue culture has proven the extended viability of tissue cells outside of the body and provided an unmatched window into the life of the cell.[1]

In 1931 Carrel published an article in the journal *Science* titled "The New Cytology," in which he complained that the conception of cells in the cytology and histology of the time was one-sidedly morphological and paid not enough attention to the physiological aspect of cells. The common practice of regarding cells as inert building stones robbed them of their vitality, he complained. "When cells are considered only as structural elements they are deprived of all the properties that make them capable of organizing as a living whole. Within the organism, they are associated according to certain laws. *Cell sociology results from these properties specific to each cell type*" (Carrel 1931, 298, emphasis added).[2] Carrel was at the forefront of tissue culture technique and made adept use of microcinematography to reveal with time-lapsed films the very active and plastic lives of cells when freed from the rigid constraints of their normal position and roles in the body or cell society. Studies conducted by Carrel and others showed that tissues and cells have potentialities hidden from common view (such as the ability to live and move with autonomy outside the body in a glass dish), their nature being more plastic than conceptions based on their structure alone, drawn from fixed and stained sections, recognized. The new cytology he called for would emphasize this dynamic and temporal aspect of cells and tissues. "Cell colonies, or organs," he wrote, "are events which progressively unfold themselves. They must be studied like history. A tissue consists of a society of complex organisms" (298). A new approach was needed, he insisted, one less reductionist, since the concepts and methods of physics and chemistry are inadequate to capture the dynamic living aspects of cell physiology. The physiological properties of cells and tissues belong not to the level of molecules and atoms but to "the supracellular order and are the expression of sociological laws" (303).

Carrel made it clear that this new way of looking at cells had significant implications for the future direction of research.

> In the development of the new cytology, as in the development of every new science, the conception is more important than the method. Techniques are only the servants of ideas . . . A method is an instrument which finds only that which is being sought. The new cytology is considering cells and tissues, not only as elements of the dead body, but as living beings which are themselves parts of organisms of a more complex order (Carrel 1931, 303).

What was new in Carrel's new cytology was not so much that it approached the cell as an elementary organism, but that it approached it as an intrinsically social one. And while its method did involve the new technique of tissue culture, it was the conception of cells as social beings with rich and interesting lives worthy of careful consideration, and the very conception that there could be such a thing as a cell sociology that was truly novel.

3. Michael Abercrombie, pioneer "ethologist" of cells, and further development of a "cell sociological" approach

By the 1940s, tissue culture had developed into a major form of biomedical technology. Researchers applied it to the study of how viruses grow in human cells in vitro, and by 1952 Jonas Salk had produced a vaccine for polio. As Hannah Landecker explains, through these developments, biomedical researchers interested in human health and disease were able to substitute the whole organism for its parts (cells), and in the process living cells became "virus-producing factories," and, more generally, cells in culture became technologies (Landecker 2007). But not all scientists at this time regarded cell culture primarily as a technology for such applied ends. In the 1950s Michael Abercrombie (1912–79), a developmental biologist at the University College of London, and his colleague Joan Heaysman, published a series of influential articles on the "social behavior of cells in tissue culture." Their approach to the study of cells in culture was much more consistent with Carrel's vision of a new cytology as cell sociology. The first paper of that series began with these words:

> A tissue culture is often referred to as a colony of cells, thereby implying that a cell can in some circumstances be regarded as a *social organism*. Yet this idea seems to have had surprisingly little effect on methods of investigation. No systematic attempt, so far as we are aware, has been made to apply to *cellular societies* the analytical approach which has been so successful in the study of animal social interaction: that is, *to take the cell as a unit and to investigate how its behavior is influenced by other cells* (Abercrombie and Heaysman 1953, 111) (emphases added).

Abercrombie noted that others had made important observations of the behavior of cells singly and in populations, but what made his studies unique was that he was "the first to develop cell behavior as a rigorous quantitative science" (Dunn and Jones 1998, 124). For his career-long efforts to carry out this approach, his colleague Sir Peter Medawar referred to Abercrombie as "the pioneer ethologist of cells" (Medawar 1980).[3] Abercrombie's most significant discovery involved the phenomenon of "contact inhibition." In investigations of chicken heart fibroblast cells in culture Abercrombie and his colleagues observed that when the leading edges of two motile cells crawling on the glass surface meet, they normally stop momentarily, and after a time move away from one another (Abercrombie and Heaysman 1953; 1954a). Abercrombie also observed that some malignant tumor cells do not display contact inhibition; they crawl over other cells and one another so that populations of malignant cells grow in heaps rather than the monolayered epithelial-like arrangement typical of normal cell populations (Abercrombie and Heaysman 1954b; Abercrombie, Heaysman, and Karthauser 1957; Abercrombie and Ambrose 1958). In this sense, the behavior of cancer cells is antisocial on both the micro (cellular) and macro (organismal) scales.

Explicit use of the term "cell sociology," however, appears to have been relatively rare, turning up in an unsystematic and disconnected fashion through the middle of the twentieth century. The French pathologist and tissue culturist Albert Policard (1881–1972) made casual mention of the idea in a popular work published in 1964, *Cellules Vivantes et Populations Cellulaires* (Policard 1964). Being a brief introduction to the contributions of molecular biology to the study of cells in isolation and as members of "les populations cellulaires," e.g., our own bodies, Policard described attempts to understand the laws governing their function and behavior as "problèmes de sociologie cellulaire" (143). The French historian and philosopher of biology Georges Canguilhem credited Policard, in conjunction with the Danish cytologist Albert Fischer (1891–1956), with having discovered that animal cells proliferate in culture only in the company of a minimal quantity of other cells (Canguilhem 2008, 45). A solitary cell or too small a number of cells soon die; but given enough of them, they will survive and begin to divide.[4] Canguilhem does not give references, and I have not been able to find a paper by Policard on the subject, but Fischer did publish a paper in 1923 on "The Relation of Cell Crowding to Tissue Growth in Vitro" (Fischer 1923). Fischer reported that experimental attempts to get an isolated fibroblast cell to divide and proliferate were unsuccessful and that proliferation was successful only "when a number of cells were in close contact in a culture" (669). When only a few scattered cells were transplanted into a culture dish, not only did they not grow, but they were seen to degenerate and die. This observation may be regarded as an example of what the English developmental biologist John Gurdon later called a "community effect." Gurdon described how in a vertebrate embryo, cells undergoing an inductive influence to become muscle tissue are dependent on close contact with other neighboring cells differentiating in the same way and at the same time (Gurdon 1988). (We will return to this important topic of a community effect.) Fischer elsewhere discussed the significance for development and physiology of the "strong social forces" and communication among cells, suggesting that "cytoplasmic bridges from cell to cell may be a physiological system . . . which provide[s] for the maintenance of a strong social system among the cell elements" (Fischer and Jensen 1946, 227, 228).[5]

The phrase "cell sociology" also occurs in a 1965 paper by the Danish biochemist Herman M. Kalckar (1908–91). In "Galactose Metabolism and Cell 'Sociology'" (Kalckar 1965) Kalckar discussed cell surface receptor molecules in bacteria and their potential role in what he called the "social characteristics" of bacteria *vis à vis* phages (viruses) and other bacteria. Kalckar mentions how bacteria with mutations for various sugar-catalyzing enzymes (epimerase particularly) exhibited altered colony morphology ("cell sociological patterns") or susceptibility to infection by phage-virus. Unable to carry out the requisite metabolism of sugars like galactose, these bacterial cells had altered cell surface receptor structure, which Kalckar suggested affected their "social relations" with other bacterial cells and viruses. Interesting as this was for microbiology, Kalckar saw a potential significance for understanding what he called the cell sociology of abnormal tissue development in higher animals such as humans. Kalckar had found that some tumor cells are highly defective in the metabolism of galactose, which is a regular component of organ-specific surface antigens of mammalian cells. This led him to propose that "conformational changes of glycoproteins may well be crucial for a variety of cell social characteristics" as the "primary factor in growth regulation may well depend on a specific contact between cells" (311–12). Noting Abercrombie's finding that malignant tumor cells lack contact inhibition, Kalckar rather presciently opined that "aspects concerning cell surface recognition patterns and their possible role in controlled and uncontrolled growth may well pose problems of particular relevance to an understanding of cell population dynamics in higher organisms" (312). This was an early contribution to the understanding of cell communication and signal reception, a field of inquiry Kalckar described as "Ektobiology" (Kennedy 1996, 158). Growing recognition from the 1960s on of the ubiquity and importance of cell-cell communication (through direct contact between cell surface receptors or via diffusible chemical molecules) would reshape cellular and organismal biology and reinforce the image of the cell as a social organism.

The connection between viruses and alteration of cell behavior was followed up by the cell culturist and cancer researcher M. G. P. Stoker in his 1971 Leeuwenhoek Lecture "Tumour Viruses and the Sociology of Fibroblasts." Stoker explained how tumor-inducing viruses reduce topoinhibition (a growth cycle variant of the motility inhibition by cell contact described earlier by Abercrombie) and modify the cell surface structure of mammalian fibroblast cells in culture. In a discussion of the "social interactions" between normal cells, Stoker described cancer cells as "asocial" (Stoker 1972, 9, 16), and in reference to Abercrombie's discovery that the movement and proliferation of a transformed cell is not inhibited by contact with other cells, he said of the cancer cell that it behaves "as though it were alone, even when it is not" (9).[6]

For those trained as experimental embryologists, observation of individual cell behavior *in vitro* offered some insight into how cells behave *in vivo* as members of a "cellular society" undergoing the highly ordered and regular processes of development and morphogenesis. With the aid of cell culture and microcinematography, biologists were able to observe living cells in ways not possible by means of the standard techniques of the old cytology, i.e., by observing slides of fixed and stained dead material. As the embryologist D. A. Ede at the University of Glasgow explained, cells in culture appeared as "semi-autonomous organisms" crawling about and interacting like "little amoeba-like creatures" (Ede 1972, 165). They do not, however, behave as though they are entirely autonomous creatures that fail to recognize one another (though this does describe to a degree the behavior of cancer cells). In fact, as Ede noted, cells of particular tissue types display what he called specific forms of "elementary social behavior." Ede used a striking analogy to illustrate how tissue culture permitted an intimate glimpse into the secret of morphogenetic coordination:

> If we looked down from a helicopter onto a swimming pool and saw on the surface a large living multicoloured object which appeared first round, then star-shaped, then long and thin, with changing combinations of colours, we should be puzzled, but if we approached closer and found we were watching a water ballet with young ladies in brightly coloured swimsuits swimming about, joining hands and touching toes in various complicated sequences and arrangements our puzzlement might be replaced by an interest in what ingenious rules they followed to enable each one to play her part in making these elaborate patterns without being able to see beyond the arms and legs of the two or three girls round about her (Ede 1972, 165).

Like the synchronized movements of the water ballet, morphogenesis involves "a population of cells, each of which behaves as an individual in a crowd, but as an individual conforming to rules by responding in ways which produces results which are most easily recognized in the form and pattern of the whole" (173). And how do the cells manage to coordinate their behavior so as to function as an orderly population? They communicate with one another.

4. Developmental biology and cell communication

One of the most significant developments in the life sciences of the last half-century is the field of cell communication. That the cells of a devel-

oping embryo communicate with one another by some means was sus-
pected at least as early as 1906 (see Shearer 1906).[7] It is now a central
thesis of modern biology that cells generally, far from being mute building stones, are in constant communication with one another and with
their environment. This holds regardless of whether they are independent, free-living cells, such as bacteria or amoebae, or tightly integrated
tissue cells in our own bodies. Biologists now regularly speak of how
cells "talk" to one another in a language of molecular signals (cf. Bassler
and Losick 2006; Winans 2002; Niehoff 2005). This rich new field of
study comprises an important amendment to the classical cell theory
and provides extra force for the conception of cells as social beings, for
communication is an essentially social activity.[8]

Cell communication is what makes possible the development and
normal function of animals, plants, and other multicellular organisms
and cell-colonies. Without communication between the cells, there
could be no coordination of individual cell activities. At best, all that
would be possible would be a heap of independently functioning cells,
with no coordinated division of physiological labor, no differentiation
of cells into specialized tissues and organs, and no pulling together for
a higher common purpose.

Cells communicate with one another by means of electrical, chemical,
and mechanical signals. Signals may be passed between cells in direct
contact by the physical forces of pressure, shears, and twists (mechanical signaling), by the transmission of charged ions through interconnecting gap junctions (juxtacrine signaling), across synaptic clefts between nerve cells (neurotransmission), locally among neighboring cells
(paracrine signals), or across great distances by hormonal messengers
carried through the blood stream (endocrine signaling) (Alberts et al.,
2008, 879ff.). By such means cells communicate to one another information about the external environment and their own internal states,
and in so doing influence one another's behavior. Cell communication is
vital to the developmental processes whereby a new multicellular organism is created from a fertilized egg cell. As the cells of the original ovum
divide, signals passed from one to another trigger the transcription (or
repression) of specific gene sequences leading to differential protein synthesis, resulting finally in the differentiation of initially similar cells into
all the specialized tissue and organ systems of the adult organism.

Developmental biologists had long pondered how the various differentiated regions of the embryo (germ layers, etc.) arise. With the discovery that certain key regions of the developing embryo have the ability to
induce changes in other regions (for instance, Hans Spemann's experi-

ments at the turn of the twentieth century on the inductive influence of the optic cup tissue to cause ectoderm at any part of the embryo to develop into eye lens tissue), embryologists knew that some form of communication of causal influence is at work. Earlier experimental interventions like those performed by Haeckel and Hans Driesch revealed the surprising regulatory and regenerative capacities of many (though not all) species of embryo to withstand injury and reorganization of their cellular components. Embryos at the two-cell stage were cleaved in half to see how far and how normally the separated blastomeres would develop; and fragments of later-stage embryos were excised and grafted into different locations in the same or different embryos. Remarkably, many of these disruptions either had little ultimate effect on the normal development of the embryo, or showed that there were distinct modules or regions of cells within the developing embryo that could be transferred from one location to another to create a predictable result in an unusual place (e.g., the growth of a limb or an eye lens where one wouldn't normally appear), or in the case of the "organizer" (a small region of tissue from the dorsal lip of the blastopore discovered by Spemann and his assistant Hilde Mangold), the ability to induce an entire conjoined twin organism.

One attempt to explain how specific regions of the embryo could control the future development of other adjacent regions involved the concept of a morphogenetic field, using an analogy with field theories in physics (Haraway 2004). The idea that a developing embryo consists of distinct fields or regions that organize the behavior and patterns of cells within their boundary was introduced in the 1920s by Alexander Gurwitsch (1874–1954) and then taken up by Paul Weiss (1898–1989) to explain limb regeneration in amphibians. The concept of fields was popular in late-nineteenth and early-twentieth-century physics and provided a contrast to the equally influential atom concept. The dispute between advocates of the cell and organism standpoint (discussed in chapter 1) can be understood against the backdrop of these atomistic and field models. For Weiss, an organicist and early pioneer of systems biology who rejected Jacques Loeb's chemicophysical attempt to reduce organic properties to the inorganic, the field represented a hierarchical level of organization and causal influence acting above that of individual cells. Beginning in the late 1960s, Lewis Wolpert amended the morphogenetic field idea to include an element of polarity and direction so that it could carry what he called "positional information." As a gradient of a chemical signal (a morphogen, in the language of the reaction-diffusion model developed in 1952 by Alan Turing) dissipated through a region of tissue, individual cells would receive information about their relative position within that field and be appropriately shuttled along particular developmental pathways. Wolpert's formal and abstract "French Flag Model" was intended to show the way from a mass of experimental data toward a general theory of pattern formation in development.[9] The morphogenetic field theory enjoyed wide popularity, despite recognized inadequacies: for instance, the implausibility that molecular morphogens could randomly dissipate through an embryo carved up into cellularized compartments, and the lack of any clear idea how exactly a cell reads its position within the purported morphogenetic field.

5.1 Rosine Chandebois and Cell Sociology

Among those critical of the morphological field model was the French embryologist Rosine Chandebois (1928–), who in the 1970s published a series of papers criticizing the idea that development relies on the causal influence of a supracellular field. She proposed instead that the interactions of individual cells were sufficient to account for the emergence of differentiated form and pattern. She called her theory "Cell Sociology."[10]

Chandebois (Emerita Professor of Embryology at the Université de Provence Aix-Marseille I), did her PhD work with the Dutch embryologist Pieter D. Nieuwkoop (1917–96) at the University of Utrecht. Chandebois's experimental investigations of regeneration in planaria convinced her that the morphogenetic field concept was inadequate and resulted in a 1976 monograph *Morphogénétique des animaux pluricellulaires* (Maloine, Paris). In the same year, she published the first of a series of papers laying out her cell sociology approach to understanding development (Chandebois 1976).

Chandebois's level of analysis was situated midway between a supracellular field and the cell as an isolated individual. As the name might suggest, the focus of cell sociology is the social interactions occurring between individual cells, both those interactions among cells *within* a specific group and interactions occurring *between* groups of dissimilar cells. As an embryo develops, groups of cells begin to differentiate together to become distinct types of cells and tissues. Whereas earlier embryologists had started from the phenomenon of induction to devise their theories (that is, the effect one group of cells has on another group), Chandebois began with the phenomenon of self-differentiation or what she preferred to call *autonomous progression of differentiation*. This refers to the ability of a group of cells removed from an embryo and placed in a lab dish to continue to develop as a particular structure *in vitro* resembling a tissue or organ and in roughly the same period of time as it would in its original *in vivo* surroundings. Because in this instance the environment in which the cell group is situated (in the presence of other cell groups or not) has no significant effect on its ability to differentiate to a particular stage, Chandebois argued that development is in its basic form an automatic or autonomous process.[11] But it is essential for autonomous progression to occur that there be a sufficient number of cells. This should call to mind the very similar observations made by Fischer in the 1920s on the density-dependence of cell growth. Chandebois noted how a single cell explanted on its own into a petri dish will dedifferentiate into a more primordial stem cell-like state, as will the cells of a too-small embryo fragment. Groups of cells therefore have collective properties not seen in the individual cells alone. Critical of the reductionist emphasis among molecular biologists on genes and of metaphorical theorizing about genetic "programs" for development, Chandebois insisted that the phenomenon of autonomous progression showed that if any such program existed, it was not contained in an individual cell or in its constituent genes. In development, cells do not behave as autonomous units each running its own computer program; rather she urged that development is a social phenomenon.[12] As Chandebois explained:

> A cell population is therefore much more akin to a human society than to a network of automata. A cell in isolation can neither maintain its activities unchanged nor respond to a stimulus that could transform *a group of cells of the same type*. In other words, cell individuality is not based on individual memory alone: we have to do, to different degrees in different tissues, with a "group effect" implying a "collective memory." And it is this aspect of the social behavior of cells that underlies the phenomenon of progression that we call development. Its course is in many respects comparable to the history of a civilization. And its study must be pursued from the viewpoint of a sociology of cells (Chandebois and Faber 1983, 25).

A cell's phenotypic features, therefore (its "individuality," in her words), are neither determined from above by some supracellular field nor are they entirely pre-formed within the cell itself. Its individuality (just like a human's personality) is an epigenetic result of the social interactions it has with those other cells with which it has the most contact in combination with its own historical genealogy or cell lineage. What Chandebois here called a group effect (in 1983) is strikingly similar to what John Gurdon would describe in 1988 as a community effect (Gurdon 1988). In both cases, communication among cells in direct contact with one another plays the vital role in these social interactions. As Chandebois wrote:

> It is well-known today that cells reciprocally exchange information *at every moment*; in other words, they reciprocally impose on each other traits of their own individuality. For this reason, cells exhibit a social behaviour . . . Consequently, one can consider that positional information is provided to any cell by its neighbours (a complement of extra-cellular information may be provided by circulating substances, especially hormones). The individuality of any cell may be viewed as the summation of different types of positional information registered by its ancestor cells (Chandebois 1977, 208–9).

One might paraphrase this idea using a familiar social analogy (think back to your own teen-age years): Who you "hang with" (within the embryo) determines the type of cell you become.[13] Chandebois referred to a group of cells with similar phenotype as a "homotypic" population (think of a distinct peer group of like-minded individuals) and described the cells in such a group as having an "elementary social behavior" —a term borrowed from Ede (1972)— meaning they share particular modes of behavior and interaction with one another.[14] Homotypic cell populations are capable of autonomous progression.

As cells in the embryo continue to divide, they aggregate into these homotypic groups of like cells, with their respective elementary social behaviors, forming what are known as cell "condensations." Condensations are the earliest stage at which tissue-specific genes are upregulated and are a preliminary step in the creation of specialized tissues (Hall and Miyake 2000). (See fig. 3.2.)

The *heterotypic* interaction between groups of dissimilar cells, of which induction is an example, results in further progressive differentiation beyond what is possible through autonomous progression alone. Chandebois, like many other biologists, used the computer metaphor of a "developmental programme" to refer to the factors possessed by a group of cells that allows them to differentiate through their mutual interactions with one another. Chandebois insisted that at the stage of condensation formation, when a cell's fate is said to be relatively de-

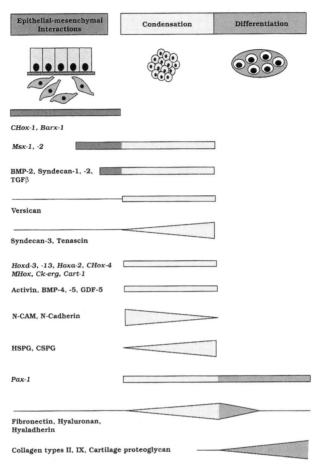

FIGURE 3.2 Cell condensation. Within-group cell interactions make possible emergent "group" or "community" effects. (Adapted from Brian K. Hall and Tsutomu Miyake, "Divide, Accumulate, Differentiate: Cell Condensation in Skeletal Development Revisited." *International Journal of Developmental Biology* 1995, 39:881–93, fig. 2, with permission of the authors and the *International Journal of Developmental Biology*.)

termined (for instance, fated to exhibit automatic progression toward a specific tissue type *in vitro*), cells possess only an elementary social behavior, which guides their (homotypic) interactions within the cell population. The remainder of the complete developmental program, she argued, is acquired in stages by heterotypic inductive interactions with other cell populations and with the broader environment (Chandebois and Faber 1983, 70). This process she compares to the trajectory of a guided missile, which receives periodically informational updates from its environment to adjust its progress toward the target

(57). Exchange of information occurs by cell contact and by hormones circulating throughout the later-stage embryo and mature animal. The complete developmental fate of no individual cell is set at the outset; and autonomous progression is a property of the cell population as a whole, not of any individual cell. The fact that autonomous progression ceases with disruption of homotypic cell interactions, Chandebois argues, "shows that the activity of a cell population is more than the sum of the activities of the individual cells" (175). The fate of a cell's individuality (its phenotype) is not predetermined from the very beginning, therefore, but is an epigenetic result of the history of the ancestral cell-line from which it derives, in addition to its own personal history, the summation of its positional information-history within the cell society. As a result of its own individual history and that of its ancestry, each cell acquires what Chandebois calls a "cytoplasmic fund" and "cytoplasmic memory," the store of molecular factors (mRNA, growth factors, transcription factors, etc.) that play an integral role in deciding which gene sequences stored away in the nucleus get expressed and to what extent. And whereas many molecular biologists at this time construed the genetic material as being in control of the animal's development, Chandebois's thinking was more in line with the cytoplasmic theories of the first half of the twentieth century (see Sapp 1987). Her metaphors and analogies for nuclear DNA were notably different from what was then the norm. She compared DNA (the genome) to the arithmetic circuits of a computer (i.e., DNA is the hardware, not the software program); and alternatively, the nucleus she compared to a stencil shop, which makes copies of the DNA/gene stencils when it receives orders to do so by the cytoplasm (Chandebois and Faber 1983, 24).

Chandebois was critical of computer-information metaphors when they were permitted to lead research ahead of or in the absence of empirical and experimental study. But otherwise she made ample use of them herself. Her understanding of a "developmental program," however, took care to note the difference between a computer program and the sort of processes that go on in a living embryo. "Our tendency to equal living things with machines" she wrote, "leads us to think that a programme must always dictate a single and inexorable pathway for the process it controls. However, the function of this kind of [developmental] automation is essentially *to propel molecular machineries through chains of events during which they themselves continuously change*" (183). If it may be said that the developing embryo operates according to some regular mechanisms, it is not the case for Chandebois that the embryo is simply a machine. In this way, she may be described as pursuing the methodology of explanatory mechanism without being a philosophical mechanist (see chapter 2). In fact, Chandebois seems firmly rooted in the holist and organicist philosophy.[15]

Chandebois's preference in metaphors also tends toward a more egalitarian conception of causal influence than the genetic determinism standard for the time.

> We have compared the pluricellular animal to a human society. In such a society each individual acquires knowledge from his contemporaries, which then comes to fruition thanks to the heritage he has received from his ancestors—in this way the individual is integrated into a particular civilization and participates in its further progress. It must be said that the present day concepts of morphogenesis conjure up an entirely different sort of society, which one could meaningfully compare to a society ruled in absolute, totalitarian fashion. All individuals are alienated from themselves and placed in a system set up nobody knows precisely how or by whom, a system that at all times forces an individuality upon them and thereby fixes in minute detail the norms of the society and the course of its future progress (1983, 169).

Similar concerns have been expressed about the metaphor of DNA as "master molecule" by architects of the molecular biological revolution, for instance the ciliate geneticist David Nanney (Nanney 1989; Keller 2002, 150–51). More social-political overtones are evident when Chandebois writes that autonomous progression, as a community effect, is an example of "mutual aid" among cells (Chandebois 1977, 212; Chandebois 1980, 3; Chandebois and Faber 1983, 56).

Cell sociology clearly emphasizes the principles of holism and emergence. This is evident in Chandebois's emphasis of the point that "the elimination of homotypic cell interactions stops autonomous progression . . . [and] shows that the activity of a cell population is more than the sum of the activities of the individual cells" (Chandebois and Faber 1983, 175). A similar holism is echoed today by other developmental biologists (see Gilbert and Sarkar 2000); and it is not so surprising to see that their choice of language also draws upon social metaphor. In "Cells in Search of Community: Critiques of Weismannism and Selectable Units in Ontogeny," Scott Gilbert writes of the difference between "dependent" (or regulative) development and mosaic development that "the whole is not only greater than the sum of its parts, the whole tells you what a part is" (Gilbert 1992, 481, and n. 4, 485).

Similarly, cancer researchers Carlos Sonnenschein and Ana Soto of the Tufts University School of Medicine have been campaigning for going on two decades now against the dominant Somatic Mutation paradigm of carcinogenesis—the thesis that cancer results from genetic mutations in one "renegade cell." Their "Tissue Organization Field Theory" maintains that non-hereditary cancers are the result of disorganization in tissue architecture resulting from a breakdown in communication between the epithelial and stromal tissue layers of which all organs are composed. Cancer is a disease of tissues, not individual cells, they argue. A tumor is, after all, a three-dimensional disorganization of tissue *in vivo*, which should give reason for pause about the prospects of the dominant approach, which attempts to understand malignant tumors through the investigation of proliferating cells in a petri dish. The book in which they lay out their developmental (as opposed to genetic) theory of carcinogenesis and the evidence for it is called *The Society of Cells: Cancer and Control of Cell Proliferation* (1999). As novel as this may sound, the authors explain that they are in fact reviving ideas expressed in the mid-twentieth century. The pioneering cancer radiotherapist Sir David Smithers launched an attack in the 1960s on what he called "Cytologism," the thesis that both normal and pathological development (cancer being one such instance) can be explained reductively in terms of the properties of autonomous cells (Smithers 1962). Smithers essentially favored the organismal perspective. An adequate account of cancer, he asserted, would require proper appreciation for the principles of organization affecting the arrangement and differentiation of cells, a "social science of the human body," which he described as lying "in wait for a name, between cytology and sociology" (498). Like many biologists who prefer a holist rather than a reductionist perspective, Sonnenschein and Soto have also argued for the significance of emergent properties in biological systems (Soto, Sonnenschein, and Miquel 2008).

Other cancer researchers, e.g., Heppner (1993) and Haroske et al. (1996), also describe cancer in terms of a society of cells and emphasize the need to concentrate on the social interactions between heterogeneous subpopulations of cells within tumors. There is also a group of cancer researchers who describe their approach to the diagnostic description of malignant cell morphology as "cellular sociology." This is, however, largely unrelated to Chandebois's theory of *cell* sociology. This group uses computer graphics and geometrical models to better describe the spatial organization of tumor cells and tissues. See, for instance,

Marcelpoil and Usson (1992), Marcelpoil, Beaurepaire, and Pesty (1994), or Kiss et al. (1995), who describe their approach as "cell population sociology." Zahm et al. (2007, 3), using cellular automata models based on time-lapse videomicroscopic recordings of tumor cells, define cellular sociology as "a large number of concepts that can be studied at the population level instead of the single cell level: migration, adhesion, cell-cell interaction and communication, cell-extracellular matrix interaction, spatial distribution, etc."

5.2 Cell Sociology and modularity in
evolutionary-developmental biology

While the direct influence of Chandebois's cell sociology approach has been modest, being taken up in name mostly by some cancer researchers interested in the development of tumors, it has more recently attracted the attention of the evolutionary developmental biologist Brian Hall.[16] Hall, who has written extensively on the development of cartilage and bones in vertebrates, has also taken a theoretical interest in the idea of modularity, i.e., the view that organisms consist of partially independent but interacting units within a hierarchy of levels (Gass and Bolker 2007, 260). According to Hall, cell condensations are one such module situated within a hierarchy between individual cells and differentiated tissues. Hall has recast Chandebois's idea of cell sociology in the language of modularity in "order to show how groups of cells maintain their collective identity during development and how signaling information from the group's environment is received by the constituent cells" (Hall 2003; Gass and Hall 2007, abstract).

A condensation, recall, is an aggregation or group of cells preceding differentiation into a specific type of tissue cell or structure. That condensations are capable of autonomous progression is evidence that they can be characterized as modules. These cell populations exhibit an emergent form of collective behavior that is unavailable to the constituent cells individually on their own. This emergent property Hall believes can be understood from the perspective of cell sociology.

> The multicellular module is stuck looking downward or upward for its identity: what it is made of, what it will make, but never quite what it is and does. In particular, the traditional emphasis on "cell-to-cell signaling" has not been concerned with framing information exchange in development as occurring between collectives of determined cells. Do cells signal other cells in devel-

opment? Certainly, but the consequences of that signal are very different depending on whether it is between cells within a module or between modules (Gass and Hall, 2007, 353).

Hall notes that continuous communication between cells within a module, Chandebois's homotypic interactions, is required for the module to retain its integration and functionality as a unit. Induction, a heterotypic communication between distinct cell condensations, is an example of interaction between modular units. Most cell condensations maintain coherent integration as a unit through direct cell contact (communication via gap junctions, for instance).

The neural crest, however, presents a more challenging example. This population of vertebrate postgastrulation embryo cells splits up into several streams of migrating cells to form a broad range of structures, including craniofacial skeleton, smooth muscle, and parts of the nervous system. Hall argues that it consists of four separate modules and should be recognized as a fourth germ layer (Hall 2000). Forming originally along the neural fold, the neural crest breaks up into four separate populations of cells, three of which migrate into the interior of the developing embryo. But the migration appears to be a collective population-level property: the migrating cells apparently all maintain gap junction contacts and do not move as isolated individuals, but as sheets. In fact, studies reveal that when communication links between cells are broken, apoptosis or programmed cell death quickly follows (Gass and Hall 2007, 7). What we have, therefore, are "migrating populations of neural crest cells," not "populations of migrating neural crest cells" (7), since the cells don't migrate as isolated individuals. Because the property in question (migration) is an emergent one belonging to the collective and not reducible to the individual cells, the cell sociological perspective provides the more adequate account. Hall writes that "condensations must attain a critical size and cells must interact within a condensation for the condensation phase to cease and differentiation to be initiated; it really is 'all for one and one for all'" (Hall and Miyake 2000, 145). One might adapt the African proverb about raising a child and say that to raise a differentiated cell, "It takes a village."

6. Programmed cell death or cell suicide as a social phenomenon

Throughout the nineteenth century, various researchers had observed that what appear to be otherwise healthy animal cells often die in large numbers, especially in certain stages of embryogenesis (Clarke and Clarke 2012). And as the observations by Fischer discussed earlier attest (Fischer 1923), animal cells *in vitro* soon dedifferentiate and die if isolated from other cells. This process of active cell death is today known to be under genetic regulation and is known as programmed cell death. An illustrative example of the constructive role of programmed cell death is the formation of digits (fingers and toes) in humans and other vertebrates. As our hands and feet develop in the womb, they initially assume the form of paddles, as condensations of cells within develop into the distinct bones of fingers and toes, and it is only after this that the cells making up the webbed tissue begin to die off in vast numbers to leave behind distinct fingers or toes. In ducks and other seabirds, this stage of programmed cell death is repressed. Cells undergoing this process expend their own energy to synthesize the proteins that will chop their nuclear chromatin and cell body into little bits, and for this reason this form of cell death has been described as "cell suicide." However, as is the case with many human suicides, what seems on the surface to be a purely solitary or autonomous act often proves to have social dimensions.

One of the earliest researchers to talk about this active form of cell death as analogous to suicide was the embryologist John Saunders Jr. By the time Saunders began his investigation of the topic in the 1960s, it had only just been recognized that the strictly timed death of normal and healthy cells was a crucial component of animal development and for the maintenance of tissue and organ homeostasis.[17] While studying the development of the embryo limb bud in chickens, Saunders found that a patch of cells in a region he called the "posterior necrotic zone" died off in a regular and predictable fashion to allow the further development of the wing and leg (Saunders 1966). However, these cells could be saved from their "death sentence" if they were transplanted to a different region of the embryo before a particular critical time. This suggested to Saunders that these cell deaths were the result of an interaction with other neighboring cells, and it raised for him the question whether this phenomenon were best regarded as "suicide or assassination" (608). The social context in which programmed cell death—or apoptosis, as it was later called, after the influential paper of Kerr, Wylie, and Currie (1972)—occurs was further explored in the early 1990s by the developmental neurobiologist Martin Raff. Raff described genetically programmed cell death as a form of "social control" exerted by the organism as a whole over its component cells (Raff 1992), in addition to popularizing the idea that apoptosis is an "altruistic" act on the cell's part (Raff et al. 1993, 699; Raff 1996). The rationale for this judgment is that not only does it serve a utilitarian purpose in the progressive development of the embryo, but it is a clean and socially responsible form of death—for as the cell self-destructs, it minimizes the risk of damage to its neighbors by presenting itself to them to be eaten or phagocytized in a safe and orderly fashion. Given that many mammalian cells die if isolated from their neighbors, Raff speculated that perhaps all mammalian cells *in vivo* are reliant upon the continual reception of "survival signals" from their peers to prevent activation of their inherent "suicide program" (Raff 1992; 1996; Raff et al. 1993; Raff et al. 1994; and Ishizaki et al. 1995). "Apparently," Raff explained, "the only thing our cells do on their own is kill themselves, and the only reason they normally remain alive is that other cells are constantly stimulating them to live" (Raff 1998, 121). If this is correct, it would mean that even the continued existence of a mammalian cell is the result of a group or community effect, and it would be a further point in favor of the cell sociology perspective.

7. Assessing the cell sociology metaphor

Evidently, there is much in the nature of animal development and function that lends itself to easy expression in the language of cell sociology. But is there more to the idea that an organism is a complicated society of cells than an attractive metaphor? Is it just window-dressing for the facts or does it do any real cognitive work? The cell biologist Claude Kordon expressed a common opinion when he insisted that "no metaphor is really explanatory," that metaphors only reflect "the cultural references through which we have been conditioned to decipher reality" (Kordon 1993, 96). Philosophers, however, are in disagreement about what makes for an explanation (see Woodward 2011 for a review). Why must an explanation be stated in purely literal terms? What is it that an explanation is supposed to provide? Hesse (1966; 1980) has argued that some, and Bradie (1998; 1999) that all, scientific explanations involve metaphorical redescriptions of one thing in the vocabulary of another.[18] Given the recent emphasis on the importance of *mechanistic* explanations in cell biology (Bechtel and Abrahamsen 2005; Bechtel 2006) and the centrality of metaphor to the very notion of a mechanism, one might legitimately ask whether an explanation in biology *can* be stated in purely literal terms.[19] While I think it is possible that many of what count as successful explanations could in principle be recast in strictly literal terms, I am more interested in how scientists actually proceed to develop explanations; and here metaphors play an undeniably significant role. I will return to the question whether metaphors can ever be truly explanatory in chapter 5.

For now, I will simply illustrate how choice of language can matter for understanding with the following—admittedly analogical—argument. Consider the picture in figure 3.3.

What do we have here? It looks like a nice "condensation" of people. But how should we describe it? Is it simply a group of people? Or is it perhaps a French class or maybe a glee club? The different descriptions suggest different types of interactions, communications, and behaviors among the individuals. So it matters quite a bit the language we use to describe it. Consider the question, "Do individuals create societies or do societies create individuals?" If we cast the question in terms of individuals, it may seem quite plausible to agree with Margaret Thatcher that there is no society, only individuals. But if we ask the question this way: "Do citizens create societies or do societies create citizens?" it now seems much more plausible to say that societies have a creative and causal influence, because to speak of citizens is to speak of a social category. Citizens are not just individuals, they're *differentiated* individuals. And likewise, to speak of differentiated cells is to talk of

FIGURE 3.3 A "condensation" of people: Glee club or philosophy class? The behavior of an individual is context-specific and influenced by the nature of the group within which it finds itself, and consequently the label used matters. Similarly for cells. (Photo by Graham Iddon.)

a group/community/social phenomenon. That is the key insight of the cell sociological approach. It is to recognize that in order to understand fully how multicellular organisms and their component cells are created and able to maintain function, one must see them, not just as groups of cells or genes or molecules, but as social phenomena with emergent properties made possible by the interaction between units of different levels of organization (from molecules, cells, cell condensations, tissues, organs, organ systems, to organisms and environment). So if we recall the phrase attributed to the nineteenth-century botanist Anton de Bary that "the plant makes cells, the cells don't make the plant," we might wish to agree that indeed the organism makes *differentiated* cells, while still recognizing the role of *individual* cells in composing and constructing the *organism*. For cells do not stand in the analogical relation to the body as bricks to a building, but as citizens to a society. Just as it takes a society to create the special category of individuals known as citizens (or students, teachers, or scientists), so it takes a society or community of cells to create specialized types of cells. Cells may be parts of organisms, but they are not parts in the same way as grains of sand are parts of a sand heap, or Lego blocks are parts of a Lego-house, or even computers are parts of a computer network. Unlike these spatially and temporally autonomous units, cells are transformed by their social interactions; they become different types of thing, they become specialized or differentiated cells.

The sociological perspective may also help us to understand how a higher-level sort of individual emerges in evolutionary time, as individual cells can only achieve what may amount to an increase in fitness as component parts of a larger group-entity. In this way, a new level of *social* agency (the group or community effect) emerges, making possible the transition to a new level of individuality, at the tissue, organ, and ultimately organismal levels. Proper understanding of cellular conditions, such as development, cancer, and evolution, requires paying attention to cell-cell interactions and tissue organization, and talk of a sociology of cells is a useful heuristic emphasizing the importance of complex and structured social (communicative) interactions among populations and subpopulations of cells (cf. Ratzke and Gore 2015; Reynolds 2017; Fantuzzi 2017).

While the sociological perspective allows us to recognize the significance of "social" phenomena in development, such as the reliance of autonomous progression or continued cell survival on what Chandebois and Gurdon called a group, population, or community effect, it seems likely, however, that such phenomena will receive an explanation in some molecular-mechanistic fashion, without necessarily being replaced by it. For instance, when Hamid Bolouri and Eric Davidson sketched an explanation of the community effect by means of the Gene-Regulatory Network analysis approach, they wrote: "The structure of the underlying gene *regulatory network* (GRN) *subcircuitry* explains the genomically *wired mechanism* by which community effect *signaling* is linked to the continuing transcriptional generation of the territorial regulatory state" (Bolouri and Davidson 2010, abstract. Emphases added to highlight key metaphorical concepts). Foreseeing an expansion of this sort of explanation, they say, "As the structure/function relations of developmental gene regulatory networks (GRNs) have come into focus, the genomically encoded wiring that causally underlies many aspects of developmental phenomenology are becoming resolved" (170). It seems likely that such reductionist and mechanical accounts of development will not entirely supersede or replace the more holistic sociological descriptions, as they seem to fulfill different but complementary roles. Just as one might have two different approaches to the study of cell phones: one focused on the electronic mechanisms of how they work, the other on the social dimensions of how they are used by different segments of society and the implications of their use for that society's future development.[20] It is also worth noting, as Baetu (2012a) has recently argued, that one should not assume that because cells and cell populations exhibit emergent properties this creates an argument in favor of antireductionism. In fact, Baetu (2014) makes the case that understanding of complex biological phenomena requires a mosaic of disparate models and techniques.

Some might also worry that cell sociology involves the anthropomorphic ascription of intentions or purposes to cells and thereby leads to "Darwinian paranoia" (Godfrey-Smith 2009, 142ff.), i.e., the fear that our genes or cells have their own goals and strategies that are at cross-purposes to our own as socially and ethically responsible human beings. Should talk about cell sociology raise such concerns? I would argue not. Cell sociology simply highlights the fact that groups of cells behave differently and have different capacities than individual cells on their own. When Matthias Schleiden remarked in 1838 that "each cell leads a double life: an independent one, pertaining to its own development alone; and another incidental, insofar as it has become an integral part of a plant [or animal]" (Schleiden [1838] 1847, 231–32), he placed his finger on why our thinking about cells is so complicated. For cells may be at once a whole and a part; and as a consequence, the question of how to talk about cells has always been at least as important as the problem of how to observe and to physically study them. Attention to the social interactions of cells reveals how cells manage to lead these double lives.

Conclusion

In her book *The Century of the Gene*, Evelyn Fox Keller reflected on the dominant metaphors of molecular biology and the guiding influence of the gene-centered approach on the understanding of development and function of living organisms in the twentieth century. The reductionist focus of the genetic paradigm has resulted in a better appreciation of the complexity of cellular and organismal biology, which has in turn led to recognition of its own limitations and inadequacies. "But these very advances," Keller writes, "will necessitate the introduction of other concepts, other terms, and other ways of thinking about biological organization, thereby inevitably loosening the grip that genes have had on the imagination of life scientists these many decades" (Keller 2000, 147–48).

In this postgenomics era, more scientists are calling for an alternative conception of living organisms that emphasizes better their nature as multilevel systems of interacting modules, no one of which can be identified as the most significant or master agent, each playing a role of equal importance for a complete understanding of how life works. There is growing recognition that the major transitions in evolution, from replicating molecules through prokaryotic cells to multicellular eukaryote organisms, is a "social process" that proceeds by constructing "new levels of organism out of separate individuals" (Queller and Strassman 2009, 3143; see Reynolds 2017.) As the authors of an essay in *Nature* wrote at the eve of the new millennium: "Cell biology is in transition from a science that was preoccupied with assigning functions to individual proteins or genes, to one that is now trying to cope with the complex sets of molecules that interact to form functional modules" (Hartwell et al. 1999, C52). The authors of a more recent review essay in *Nature* likewise remark that "proteomic studies have revealed detailed lists of the proteins present in a cell. Comparatively little is known, however, about how these proteins interact and are spatially arranged within the "functional modules" of the cell: that is, the "molecular sociology" of the cell" (Robinson, Sali, and Baumeister 2007, 973). So it seems that the sociological approach is even finding adherents among molecular biologists as well as developmental and cancer biologists.

Even in the heart of one of the most reductionist and mechanistic fields of inquiry, cell signaling, one finds expression of the social perspective. As the lead editorial for the inaugural issue of the journal *Cell Communication and Signaling* states: "It is now well established that cells do not behave as selfish entities but rather tend to form <<microsocieties>> whose proper functioning requires a precise coordination of emission and reception of signals" (Perbal 2003, 1). Far from being incompatible, the mechanistic and social perspectives in cell biology are complementary stances biologists assume when and as required. For instance, neurobiologist Julia Kaltschmidt and geneticist Alfonso Arias write that "in essence development is about cells and their social life, about how their interactions result in diverse tissues and structures such as arms, eyes, or lungs and about instruction manuals that arrange these in space to generate distinctive organisms" (Kaltschmidt and Arias 2002, 316). But when discussion turns to the details of these social interactions, the authors speak of the cell as a "network of machines capable of emitting, receiving and processing information" (318). A similar willingness to employ both sorts of metaphors is evident in the work of Rosine Chandebois, who despite being cautious of uncritical comparisons between cells and computers, made ample use of the idea of a developmental program herself (cf. especially Chandebois and Faber 1983, 4–5).

This pragmatic shift from one perspective to the other is not entirely new, of course. Biologists have long attempted to understand the living organism as a kind of machine. Part of the novelty of the cell theory has always been the suggestion that humans and other organisms are made of little modules (cells), which behave like little organisms in their own right (communicating with one another, etc.). This perspective prompts a wealth of questions about the structural and functional relationships between these cellular modules for which social metaphors and analogies can provide revealing insights, but when the question is *how* do cells and other organisms work?, the reply "Like machines" has always had great attraction. Perhaps we could say that the organismal-social perspective helps us with *knowing that*, where the machine perspective helps us with *knowing how*. Knowing *that* cells behave in such-and-such a way is not equivalent to knowing *how* they manage it, where machine metaphors play a crucial role in helping to construct explanations in terms of mechanisms. Consider the example of active cell death, where its metaphorical description as a form of suicide serving altruistic ends, but also as a form of execution in the interest of social control, helped to inspire interest in its occurrence and to situate it in a broader social or organismal context. But in order to understand the mechanism of how the process is initiated and executed, scientists made productive use of the computer metaphor of programmed cell death (Reynolds 2014).

In her exposition of explanations of animal development, Cor van der Weele identified three competing perspectives: genetic, structuralist, and constructionist, each of which is based on a particular metaphor: program, field, and construction, respectively (Van der Weele 1999). Though I have focused on two slightly different metaphorical perspectives here (the social-organismal and the machine-mechanistic), I share her opinion that no single approach can be complete, as every "perspective" is selective. Or, using an alternative metaphor, we might say these perspectives are different tools, and just as there is no universal tool that is optimal for every task so we should not expect to find one perspective or approach that is universally or canonically correct. At best, something like Denis Noble's "middle-out" approach might be urged to integrate all the various levels of organization necessary to understand development and organismal function (Noble 2006). Admittedly, nothing I have said here demonstrates that the metaphor of cell sociology or its close cognates have ever played an ineliminable cognitive role in either the process of discovering some important biological fact or explaining one. It may be that their utility comes after the fact, as it were, and is of a purely heuristic nature in that they provide a handy language with which to talk about the facts. But that itself would be an important lesson for those who seek to understand how successful sciences operate. And by encouraging scientists to think about cells and organisms in this particular way (as a complement to the machine paradigm), they may be more likely to recognize emergent features and community effects resulting from the social interactions among cells.

Machine metaphors have undoubtedly proven useful when dealing with relatively static or constant cause-effect relations in which the parts undergo no significant historical changes that fundamentally affect their basic properties. But where the parts or the whole do have a historical dimension, then social-agential metaphors may be more useful. Cells, unlike machines, building stones, or atoms, have "stories" about how they got that way, they have histories;[21] and, as it turns out, many intracellular parts are less machine-like in this regard too—their histories make a difference to their morphology and function—and, in further contrast to standardized machine-parts (think BioBricks™), sometimes a large stochastic element is important, as is becoming clearer in the case of "intrinsically disordered proteins," which are turning out to have important roles in intracellular signaling pathways. These developments in how cell and molecular biologists think about the complex dynamics of cell physiology and behavior are the subject of the next chapter.

4 Cell Signaling: The cell as electronic computer

If the seventeenth and early eighteenth centuries are the age of clocks, and the later eighteenth and the nineteenth centuries constitute the age of steam engines, the present time is the age of communication and control.

Norbert Wiener (1961, 31)

1. Introduction

We saw in the last chapter that biologists describe the complex developmental and physiological interactions of cells in terms of cell-cell communication, and that some think about these interactions in terms of the metaphor of cell sociology. Much of this discussion of cell-cell interaction is descriptive in that its primary concern is to catalogue the facts about cell behavior. But as we saw in the second chapter, when scientists attempt to articulate causal explanations of *how* cells manage to communicate with one another, they often rely on the language of machines and engineering metaphors. This chapter begins with a brief history of scientific accounts of cell-to-cell communication and of the molecular processes involved in intracellular signaling, i.e., the events taking place inside the cell when it sends and receives messages from other cells or its environment. The theory of intracellular signaling in particular has been dominated by metaphorical language borrowed from electrical engineering, so rather than being anthropomorphic like the language of cell sociology, the scientific discourse of signal transduction is quite strikingly *technomorphic*. The chief background metaphor under consideration in this chapter will be THE CELL IS A COMPUTER or similar electronic device.

However, scientists working on the problem of understanding how an external signal is interpreted by a cell also frequently use agential metaphors to describe the activity of proteins and other components within the cell. Recently there have been criticisms of the machine- and computer-engineering-dominated conceptions of the events involved in the intracellular processing of external signals or stimuli. These alternative proposals, arising from increased attention to the complex temporal dynamics of what were originally conceived to be signaling pathways and genetic circuits, tend also to make greater use of agential metaphors. As a consequence, the sociological perspective has revealed itself to be useful for thinking about some aspects of cellular physiology. The molecular components of signaling pathways are now described by some biologists as dynamic and flexible agents that "cooperate" to construct and deconstruct various temporary signaling pathways, resembling more a team of workers engaged in a suite of group activities than a static and hard-wired computer circuit board. But here again the trend seems not to be that one set of metaphors is entirely replaced with another, but that the two are serving complementary functions. The implications of this trend of dual metaphor use for a philosophical account of science will be treated more fully in chapter 6.

2. Cell communication: coordination and control
of cell-parts in the organism as a whole

According to the theory of the cell-state (see chapter 1), complex multicellular organisms like us are analogous to cities or states composed of many little individual cells, each of which are like the citizens among whom are divided the various tasks required to keep the social polity running. A sophisticated modern state requires many different professions and trades: doctors, lawyers, educators, construction workers, plumbers, electricians, agriculturalists, waste management technicians. In a human society, communication is essential to instigating and maintaining this coordinated division of labor. Oddly enough, advocates of the cell standpoint in the nineteenth century were rather quiet about the essential function of communication within the cell-state. Perhaps because they were so busy defending the legitimacy of the cell as a scientific concept, they tended to emphasize (and in some cases exaggerate) the autonomy of the cell in the larger organism itself, and this may have led them to ignore or to overlook the significance of communication between cells. It is true that by mid-nineteenth century the nervous system was commonly compared to a system of telegraph wires carrying electrical signals throughout the animal body (Otis 2001), but that all cells in the body might be communicating with one another was hardly thinkable, it seems. In the first and second editions of *The Cell in Development and Inheritance*, E. B. Wilson stated, "There is at present no biological question of greater moment than the means by which the individual cell-activities are coördinated, and the organic unity of the body maintained; for upon this question hangs not only the problem of the transmission of acquired characters, and the nature of development, but our conception of life itself" (Wilson 1896, 41; 1900, 58). In this regard, Wilson discussed the presence of slender "protoplasmic bridges" often found between the tissue cells of plants and animals (Wilson 1900, 59–61) and conceded that evidence suggested they are probably not "merely channels of nutrition, as some authors have maintained, but paths of subtler physiological impulse" (61). Wilson never uses the term "communication" to describe the problem of "coordination" of the body's cells, and he concluded that in the absence of further investigation, "judgment should be reserved regarding the whole question of the occurrence, origin, and physiological meaning of the protoplasmic cell-bridges" (61). Six years later it could still only be hinted at that even the cells of a developing embryo might be in communication with one another by means of these slender protoplasmic bridges (Shearer 1906); and even in this case, when the author spoke of communication, he used it as a noun referring to a physical structure rather than as a verb indicating an activity. Even as late as 1924, Robert Chambers (1881–1957), an embryologist and expert on cell microdissection, doubted the existence of protoplasmic bridges between normal metazoan cells (Chambers 1924, 242–43). However, experiments performed around the turn of the twentieth century showing the disruptive effects of killing or removing early embryo cells, and the inductive influence of tissues such as Spemann's organizer on developmental processes, all pointed to the existence of some kind of coordination between the developing structures within an embryo. How this was achieved was unclear, although some chemical basis seemed likely (Armon 2012.)

It was physiologists investigating the coordination of organ functions in mature animals by means of chemicals circulating through the blood who made the first important breakthrough. This began in 1902 with the discovery of "secretin," a peptide secreted by the cells of the intestinal lining that was shown to regulate activity of the pancreas. In 1904 one of its discoverers, Ernest Starling (1866–1927), proposed that body organ function is coordinated by chemical "messengers" like secretin carried through the blood system that act as a system of communication separate and distinct from that of the nervous system. In 1905 Starling and his collaborator Sir William Bayliss (1860–1924) dubbed these chemical messengers "'hormones" (from the Greek for "I excite"). Thereby, the field of endocrinology was established as the study of the chemical hormones secreted from the body's ductless glands directly into the bloodstream (Henderson 2005). Efforts to identify other chemical messengers responsible for coordinating the body's various functions resulted in the identification and isolation of other blood-borne messengers, such as adrenaline or epinephrine (Gomperts, Kramer, and Tatham 2009, 11–12). Insulin, the hormone responsible for regulating sugar metabolism, was successfully isolated by Frederick Banting (1891–1941) and Charles Best (1899–1978) in 1922. As a result of this progress in identifying numerous hormone regulators, the anatomist Sir Arthur Keith (1866–1955) could write in 1924 a popular essay extolling the virtues of Herbert Spencer's analogy between the body politic and the body physiologic, with an updated discussion of how the society of cells is governed by a "postal system" of intercommunication using hormone messages to deliver commands to the various organs of the body (Keith 1924). Keith described how, through "the government of hormones . . . the cells or units of the human body represent an immense assemblage of conscript citizens" (8) who have no choice but to obey automatically the instructions delivered to them by the hormone messengers.[1]

3. Cell surface receptors, second messengers, and signal transduction

By the late 1930s, endocrinology was coming to be known as the science of "chemical communication" (Sinding 1996, 47). There was a general conviction among endocrinologists that the efficacy of hormones was reliant on internal cellular organization (recall the developments in chapter 2), so that effective investigation of how they regulate the metabolic activity of their target cells required study of intact cells and tissue slices, rather than the "grind-and-find" techniques of biochemistry used to investigate enzyme activity in cell-free homogenates *in vitro* (Tepperman 1988, 314). Exploration of the sequence of events occurring between the delivery of the hormone messenger and the final transformation of sugars, fatty acids, and proteins by the action of enzymes (an area of study known as intermediary metabolism) did, however, follow the biochemical techniques of cell-free solutions *in vitro* (Sinding 1996). Thus, in the early 1950s the biochemist Earl Sutherland (1915–74) was working on the role of enzymes in intermediary metabolism when he became interested in how the hormone glucagon acts on the enzyme phosphorylase in the conversion of glycogen to glucose. As a biochemist, Sutherland ignored the orthodoxy among endocrinologists and studied the hormone action of epinephrine and glucagon on phosphorylase in broken liver-cell fragments *in vitro*. Although levels of phosphorylase activity did not increase under the influence of glucagon and epinephrine in the cell-free system as was hoped, Sutherland did discover in 1956 a new factor, which turned out to be cyclic-adenosine monophosphate (cAMP). cAMP—neither an enzyme nor a hormone—increased levels of phosphorylase in the cell-free systems, but only because the solutions in question contained fragments of the cell membrane to which it is attached. Eventually it was decided that the hormone, acting as a "first messenger," activates a receptor on the cell surface, which triggers the enzyme adenylyl cyclase to convert the peptide adenosine triphosphate (ATP) to cAMP, which then acts as a "second messenger" in the cell interior (see fig. 4.1).

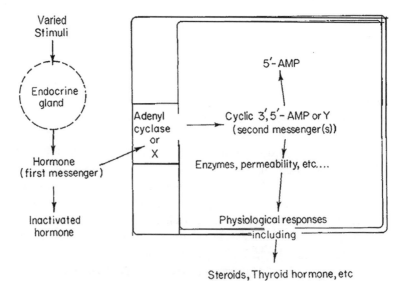

FIGURE 4.1 Schematic or "cartoon" of the second messenger model of insulin action. (From Earl W. Sutherland and G. Alan Robison, "The Role of Cyclic-3', 5'-AMP in Response to Catecholamines and other Hormones." *Pharmacological Reviews* 1966, 18(1):145–61, fig. 2, with permission of *Pharmacological Reviews*.)

It was the endocrinologist Oscar Hechter (1917–2003) who suggested that hormones were unlikely to act directly on their enzyme substrates by slipping through the cell membrane, but rather by binding to a surface receptor in a fashion similar to the action of many drugs.[2] Hechter was a close associate of Norbert Wiener (1896–1964), the chief founder of cybernetics, and consciously described hormone activity on the target cell in terms of cybernetics and information theory. For instance, writing in 1964 about Sutherland's identification of cAMP as a second messenger, Hechter said: "From a cybernetics point of view, hormone involves the transduction of environmental information, represented by the hormone signal, into the language of intracellular signals which provide information for various effector sites, giving rise to hormonal response" (quoted in Sinding 1996, 55).

The idea that the information content of a hormone signal is transduced across the cell membrane was taken up by the biochemist Martin Rodbell (1925–98), who was trying to understand the activity of insulin hormone at the cell level. Rodbell made popular the theory of "signal transduction" and in 1994 shared with Alfred Gilman (1941–2015) the Nobel Prize for physiology or medicine for his work on the role of G protein-coupled cell membrane receptors in insulin activity. In the late 1960s and early 1970s, Rodbell was pursuing Hechter and Sutherland's idea that the insulin signal (a first messenger) binds to a specific protein receptor lodged in the surface of the cell membrane and spanning the space from outside to the cell interior, when he suggested the signal is transduced or carried across the plasma membrane to a G protein complex lying on the other side.[3] The binding of the insulin molecule or "ligand" alters the shape of the associated G protein on the inside of the membrane, triggering the activity of a second messenger, further proteins and molecules, which may "amplify" the original signal and set off a "cascade" of interactions within the cell, resulting ultimately in the regulation of gene expression or other cellular response (see fig. 4.2). G-coupled protein receptors have turned out to be one of the major families of membrane receptors active in a wide array of signal transduction pathways (Gomperts, Kramer, and Tatham 2009).

Rodbell explained in his Nobel acceptance speech that at the time he was thinking about the problem of hormonal action, he was strongly influenced by Wiener's writings on cybernetics (Rodbell 1995). Wiener was interested in finding a unified account of self-correcting behavior in machines and organisms that would rely on purely mechanistic principles, or more accurately, principles of computer theory and electronics. "If the seventeenth and early eighteenth centuries are the age of

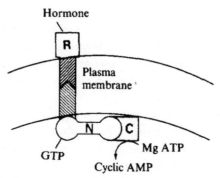

FIGURE 4.2 Model of the G-protein receptor and signal transduction. (Reprinted by permission from Macmillan Publishers Ltd. From: Martin Rodbell, "The Role of Hormone Receptors and GTP-regulatory Proteins in Membrane Transduction." *Nature* 1980, 284 (5751):17–22, fig. 1.)

clocks, and the later eighteenth and the nineteenth centuries constitute the age of steam engines, the present time is the age of communication and control," Wiener wrote (Wiener, 1961, 31). In searching for a unified account of control and communication applicable to both animals and machines, Wiener asked, "What is a machine?" To which he answered: "As the engineer would say in his jargon, a machine is a multiple-input, multiple-output transducer" (Wiener 1964, 32). As Rodbell explained in his account of how the hormone signal is transduced across the cell membrane, "The transducer is a coupling device designed to allow communication between discriminator [receptor] and amplifier [adenylyl cyclase]" (Rodbell 1995, 339; insertions mine). The cell membrane of eukaryotes is arranged as a double layer of lipid molecules with the electrically charged (hydrophilic) "heads" facing away from one another and their electrically neutral (hydrophobic) "tails" reaching toward one another. This creates an internal space through which lipophobic molecules like insulin and many other hormones cannot pass. But upon "docking" at the receptor site, the insulin ligand initiates a change in the conformational shape of the receptor, and because the receptor spans the cell membrane (seven times, in fact, like a long hairpin with multiple bends) the informational content of the message carried by the hormone messenger is carried across the membrane and converted into a different molecular effect, namely the creation of cAMP from intracellular ATP by adenylyl cyclase. The G-protein complex (consisting of three subunits: α, β, and γ) is a regulator of adenylyl cyclase activity. When insulin binds with the receptor it switches off the inhibitory effect of the associated G-protein complex, allowing adenylyl cyclase to convert ATP to the second messenger cAMP. The G-protein receptor complex acts, therefore, somewhat like a bridge across the "moat" of the cell membrane, but with the important difference that the hormone molecule or messenger itself does not pass over the bridge, only its message, which is converted or transduced in the process.

Transduction is a term originally used by electrical engineers and physical scientists to refer to the transformation of energy or information from one type to another (Gomperts, Kramer, and Tatham 2009, 1). The *Oxford English Dictionary* defines the verb transduce as "the action of leading or bringing across" and "to convey from one place to another" (quoted in Gomperts, Kramer, and Tatham 2009, xxii–xxiii). Ironically, as noted earlier, this is very close to a literal translation from Greek of the word metaphor: a device for carrying something from one place to another (cf. Gould 1995). "Transduction" had also appeared earlier as a term in microbiology (Zinder and Lederberg 1952), where it was used to refer to the transfer of genetic material from one bacterial cell to another by means of a virus or virus-like vector. The attractiveness of the transduction metaphor in this case may also be explained by the increasing penetration of information theory and cybernetics into the discourse of molecular genetics beginning in the 1950s. The information carried by a gene is then conceived as being transduced from one bacterial cell to another by means of a viral vector.[4]

Others before Rodbell had been exploring the application of cybernetic theory to cellular biology (e.g., Waddington 1957) and to cell communication in the context of developmental biology (Apter and Wolpert 1965; Apter 1966). An important and, it seems, overlooked contribution to this development was the discovery in the mid-1960s of the passage of electrical currents between cell junctions in nonneural cells by the biophysicist Werner R. Loewenstein (1926–2014), working at the time at Columbia University. In a series of papers, Loewenstein and colleagues showed that communication by electrical means between cells in direct contact through interconnecting gap junctions was possible (Loewenstein 1964). They also showed that cancerous cells—which lack the feature of contact inhibition seen in normal cells *in vitro*–are apparently incapable of electrical communication because they lack the ability to form functioning gap junctions (Loewenstein and Kanno 1966). Cancerous cells are, as it were, deaf to the messages of their fellow tissue cells telling them to maintain an orderly existence. Cancerous tumors arise, then, as a result of a breakdown in communication. The author of a review of cell communication compares Loewenstein's work on the communicative function of gap junctions to the discovery of the Rosetta Stone, because it provided an important clue to the interconnection between the evolution of early multicellular metazoans, stem cells, and cancer biology (Trosko 2011).

Loewenstein was also writing about signal transduction in nerve cells (Pacinian corpuscle mechanoreceptors) as early as 1960 (Loewenstein 1960; 1965). In fact, accounts of sensory receptor activity had been described in terms of transduction at least a decade earlier. The electrophysiologist Edward F. MacNichol (19??–2004), of the Department of Biophysics at Johns Hopkins University, began his 1956 paper "Visual Receptors as Biological Transducers" with the following statement: "Receptor organs may be considered to be transducers whereby particular forms of energy coming from the environment are, to use concepts borrowed from the communications engineering field, filtered, amplified, compressed, and encoded in a form suitable for transmission to the central nervous system" (MacNichol 1956, 34). Loewenstein (1960) attributed the demonstration of a transducer mechanism in a sensory mechanoreceptor (the muscle spindle) to Bernhard Katz (1911–2003), who in 1950 showed that a mechanical deformation (stretching) of the muscle spindle of the frog generates a local electric current in the attached sensory nerve ends (Katz 1950). Katz did not explicitly use the term transduction to describe this conversion of mechanical stimulus to electrical signal—though he did refer to an earlier "transducer effect" described by Stevens and Davis (1938) in the auditory nerve. What Loewenstein found especially noteworthy in Katz's 1950 paper was his attempt to model the biological phenomenon with a simple electrical circuit diagram, an example he followed in his own investigations and publications.

It is worth noting that Loewenstein's use of the term signal in his papers on gap junction intercellular communication was void of any specific content or physiological meaning. Loewenstein and his collaborators were experimenting with chains of interconnected epithelial cells isolated from *Drosophila* salivary glands, using microelectrodes to pass a current of electricity through one end and measuring its detection at the other. What they found was that the current flowed with little to no resistance between the cells at the sites of gap junctions, whereas there was significant resistance elsewhere on the cell membrane surface. But the signal in these experiments was without biological meaning or physiological specificity.[5] It was a signal just in the sense that any voltmeter measures—if the needle is deflected, it indicates the presence of an electrical signal, but the signal does not necessarily *mean* anything. In general, the meaning of any cell signal lies not in the chemical messenger itself (the molecule) but in its *effects* on the target cell. Appealing to the pragmatist theory of meaning developed by the American scientist-philosopher Charles S. Peirce (1834–1919), we might say that the meaning of a signal lies in its consequences (or what Peirce called its interpretants), not in anything intrinsic to the messenger molecule. The same molecule can have different effects or meanings in different types of cells and in fact in the very same cell at different times depending on the internal state of the cell.[6]

There is a general ambiguity in the concept of *signal* as it appears in discussions of cell signaling and signal transduction, an ambiguity reminiscent of the concepts of molecular *information* and of a developmental or genetic *program* as described by Evelyn Fox Keller (2002). But as Keller argued with respect to the program concept, it is from this ambiguity or "ambi-valence" that the signal concept derives its power. It is this ambiguity in meaning that facilitates the metaphorical transfer of concepts from electronics and cybernetics to cell biology. These earlier efforts to apply the concepts and techniques of electrical theory and engineering to the physiology of nerve cells may help to explain why biochemists and endocrinologists like Rodbell found it natural to use the language of signal transduction to describe what is essentially a chemical process. Cells are not, after all, hard-wired devices like transistor radios or computers. As some biochemists like to say, "Biology is nothing but applied chemistry!"[7] So why model a chemical system with electronic analogies?[8]

4. Intracellular signaling: from pathways to networks

Once light had been shed on how a hormone signal is transmitted from outside the cell across the cell membrane, the next task was to work out the ensuing set of events that result in a physiological response to the initial message or signal. Beginning in the 1970s, the complete set of molecules and events involved in a cell's response to an external stimulus was referred to as a "signaling pathway," in analogy with the earlier concepts of metabolic and developmental pathways. The signaling pathway concept has been especially influenced by metaphors and analogies from electronic engineering and cybernetic theory. As the French cell biologist Claude Kordon (1934–2008) remarked, scientific understanding of cell physiology from the nineteenth century to the present has frequently relied on metaphors drawn from human technology: from early machines, to telephone systems, to current computer systems (Kordon 1993, 95–96). By the 1990s, use of the term signal pathway had become ubiquitous in the literature. Following the

FIGURE 4.3A Visual metaphor of the cell as electronic circuit. (a) *Science* cover 31 May 2002, vol. 296, issue 5573. Image: Julie White. (Reprinted with permission from AAAS.) (b) *Science Signaling* cover 21 October 2008, vol. 1, issue 42. "The image is an artist's rendition of signaling networks rendered as electronic digital circuits and was inspired by the Research Article by Abdi *et al.*" Image: Christopher Bickel. (Reprinted with permission from AAAS).

analytical-reductionist techniques of molecular biology, the signal pathway is dissected into component signaling molecules, receptors, protein transducers, second messengers, amplifiers, effectors, and so forth. The mechanism of intracellular signaling is now widely conceptualized as consisting of circuits and programs, which scientists are busy trying to map in an effort to understand development, health, and disease in humans and other organisms (see fig. 4.3).

Early biochemical studies into cell-cell communication focused on discerning the kinetics and reaction rates between various molecules: hormones, growth factors, cytokines, etc. In the 1950s it was discovered that a class of enzymes called protein kinases activate other enzymes by transferring to them a high-energy phosphate group (typically from a molecule of ATP) in a process known as phosphorylation. By the 1960s, kinases were being described as metabolic switches with regulatory

FIGURE 4.3B (*continued*)

function over other enzymes and proteins.[9] Other protein phosphatases act as off switches by removing the phosphate and thereby shutting down the enzyme activity. One of the many effects of the second messenger cAMP was found to be the promotion of kinase activity. Protein kinases and phosphatases are often organized into a complex chain of causal interactions, whereby an initial signal can be amplified and dispersed throughout the cell in what is known as a cascade, invoking the image of a stream of water falling down a terrain and breaking into multiple streams or channels—(cell biologists often speak of events occurring upstream or downstream of some particular point in a pathway). Visual representation of the relations between the various molecules involved in a signal pathway, illustrating which ones stimulate or inhibit the levels and activity of which others, took on the form of flow charts and electronic circuit or wiring diagrams (also known as "block diagrams"— see fig. 4.4). As the term suggests, these signaling pathways were initially assumed to be more or less linear and well "insulated" from one another within the cell's internal organization so as to maximize the specificity and precision of an external signal, as one might expect to be favored by millions of years of natural selection. One complication in the strict linearity of these pathways had been recognized earlier by Hechter and other enthusiasts of the cybernetics program in the form of positive and negative feedback loops, whereby a chemical

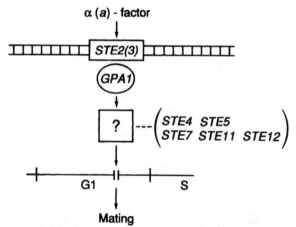

FIGURE 4.4 Early signal pathway diagram. (Republished with permission of American Society for Microbiology—Journals, from N. Nakayama, Y. Kaziro, K. Arai, and K. Matsumoto, "Role of STE Genes in the Mating Factor Signaling Pathway Mediated by GPA1 in *Saccharomyces cerevisiae*." *Molecular and Cell Biology* 1988, Sept. 8(9):3777–83, fig.3, permission conveyed through Copyright Clearance Center, Inc.)

agent could regulate the concentration and intensity of activity of itself or other molecules.

The signaling pathway metaphor is a rather natural and efficient tool for attempting to organize conceptually the sequence of biochemical events involved in the coordination and regulation of cell physiology. But let's consider some of the significant assumptions about the biological reality under investigation it introduces. First of all, signal pathway is an ontological metaphor that suggests there is an entity within the cell responsible for the activity of communicating an external signal or stimulus to the cell interior, when in fact the transmission of a stimulus into the cell is more accurately construed as a dynamic flux in the concentrations of various molecules. A literal pathway, say, one through a garden, exists as a distinct, coherent, and static entity even if and when no person or thing is traveling down it. But that is not quite the case with the chain of biochemical reactions scientists are trying to denote by means of the term intracellular signaling pathway. In the words of Tony Pawson (1952–2013), the 2008 winner of the Kyoto Prize for his work in signal transduction: "The inside of a cell is somewhat like a jigsaw puzzle, though one that is very complex because it is constantly changing shape" (Pawson 2008).

Gradually throughout the 1980s and 1990s, there was increasing recognition that many of the supposedly distinct signaling pathways interacted with one another via various second messengers and other molecules, which played a role in two or more pathways, thereby acting as a junction between them. These interactions between what were construed as distinct pathways are referred to as crosstalk, another metaphor borrowed from the field of electrical engineering, where it denotes an undesirable result of faulty design or poor insulation between neighboring wires or circuit components. Eventually, biologists started to regard this crosstalk between pathways as not necessarily a design flaw but an efficient and economical means for cells to integrate signals and obtain greater signal specificity. Consequently, the signal pathway concept has been increasingly criticized for being misleadingly simplistic and an impediment to further progress in understanding cell behavior, not to mention successful biomedical intervention in the treatment of diseases arising from altered cell communication and intracellular signaling. As a result, in the 1990s researchers began increasingly to employ the metaphor of signaling networks and to emphasize that all pathways are really parts of larger signaling networks (see figure 4.5).[10]

Equally influential as the signaling pathway metaphor is the circuit metaphor. As an example of its significance, I present a particular

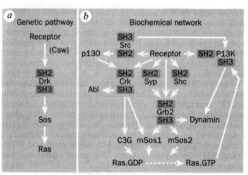

FIG. 4 Signalling through the Ras pathway. A linear pathway or m complex network? *a*, Genetic and biochemical experiments in *Dro phila* suggest a linear pathway, in which receptor PTKs couple to (the homologue of Grb2), possibly with the SH2-containing PTP Corkscrew (Csw; the homologue of mammalian Syp) acting as intermediate[125]. Drk then activates Sos. *b*, In mammalian cells complex network of interactions has been identified, involving prote that can regulate Ras, only part of which is shown here. Interacti are as discussed in the text. See Fig. 2 for details of protein structu The C-terminal proline-rich region of dynamin can bind not only G and PI3K, but also PLC-γ1. The Crk SH3 domain interacts *in vivo* v both the Ras GNRP C3G, and also with a proline-rich motif in the terminal tail of the Abl PTK[126,127]. PI3K makes multiple interactic with receptor autophosphorylation sites through its SH2 domain, v the SH3 domain of Src family kinases through a proline-rich region p85, and with GTP-bound Ras[128]. These complexes, along with p tyrosine phosphorylation, may act synergistically to regulate P activity.

FIGURE 4.5 Contrast of signal pathway with signal network model. Throughout the 1990s recognition of interaction ("cross-talk") between the components of distinct signal pathways led to increased use of the signal network metaphor. (From Tony Pawson, "Protein Modules and Signalling Networks." *Nature* 1995, 373 (6515):573–80, fig. 4, reprinted by permission from Macmillan Publishers Ltd.)

example from the area of cancer biology. In 2000, the cancer research-ers Douglas Hanahan and Robert Weinberg published their influential "Hallmarks of Cancer" paper in the journal *Cell*.[11] In this review pa-per, Hanahan and Weinberg introduced a multiple-stage model for the emergence of a malignant cancer cell through step-wise acquisition of the following capacities: self-sufficiency in growth signals, insensitiv-ity to growth-inhibitory (antigrowth) signals, evasion of programmed cell death (apoptosis), limitless replicative potential, sustained angio-genesis, and tissue invasion and metastasis. The model was articulated by means of a circuit metaphor to understand normal cell function and how defects in these regulatory circuits result in pathological activity. "Progress in dissecting signaling pathways has begun to lay out a cir-cuitry that will likely mimic electronic integrated circuits in complex-ity and finesse, where transistors are replaced by proteins (e.g., kinases and phosphatases) and the electrons by phosphates and lipids, among others" (Hanahan and Weinberg 2000, caption for fig. 2, 59). They pre-dicted that in two decades the completion of the "wiring diagrams" for every signaling pathway will reveal the complete "integrated circuit of the cell," and with this, scientists will be able to apply mathematical models to reveal how genetic lesions reprogram a normal cell to become a cancer cell (67).[12] In an update to their earlier statement, Hanahan and Weinberg (2011) add two new features (and metaphors) to their list of cancer cell hallmarks: the "reprogramming of energy metabolism" and "the evasion" of destruction by cells of the immune system.

As a further example of how biologists frequently speak of the cell simultaneously in the two disparate metaphorical vocabularies (machine and agential), the authors write: "The structure of the apoptotic *machinery* and *program*, and the *strategies* used by cancer cells to *evade* its actions, were widely appreciated by the beginning of the last decade" (Hanahan and Weinberg 2011, 650) (italics added). The cancer cell is depicted here like a secret agent attempting to outwit or deactivate a tracking device or explosive—and in fact the immune system and other subcellular components such as the tumor protein p53 ("the guardian of the genome") are frequently spoken of by Hanahan and Weinberg and others as a "surveillance system" for ensuring cell and genomic integrity.

But if the pathway metaphor is imperfect, it should come as no surprise that the circuit and computer metaphors have also faced criticism. As the authors of a recent textbook on cell signaling explain:

> Today it has become trendy to compare cells, brains, or even organisms with computers, and vice versa, to denote electronic networks capable of learning as "neural." However, this concept, although it has the advantage of clarity, rapidly leads into a cul-de-sac. Biological data processing—even if based on the same mathematical principle—is certainly more ingenious and sophisticated than today's computer technology, and it is still far from clear whether it can ever be interpreted and imitated by a technological approach. So one should treat the computer metaphor with caution, being aware of its narrow limits (Marks, Klingmüller, and Müller-Decker 2009, 2).

Even someone as fond of electronics metaphors as Dennis Bray, for whom the cell is a "robot made of biological materials" (Bray 2009, ix), concedes that there are no distinct conductive wires or traces in cell signaling pathways: "In fact the term *biochemical circuits* is flawed in several respects . . . In reality, a signal traveling through a cell is a change in the numbers of specific molecules at particular locations" (87). Like a pathway, electronic circuits are static, fixed, distinct entities (soldered or, more typically today, etched into a printed circuit board), whereas the signaling pathways of biological cells are transient, fluid systems with an important temporal dimension. Models drawn from solid state physics are therefore being applied to what is really a domain of chemistry. And as mentioned above, where electronic circuits are insulated against crosstalk, cell signaling pathways are not—they interact, providing another level of complexity. Moreover, signals within an electronic device typically carry a distinct and specific informational meaning (pressing the same button gives always the same result); whereas in biology the same messenger can have distinct effects or meanings in different cells and even in the same cell at different times. The informational content or meaning of a biochemical signal is often a function of the internal state of the cell, sensitive to the concentration of preexisting second messengers, protein kinases, transcription factors, and other molecular components resulting from the cell's history and environment. For this reason, some cell biologists refer to a temporally sensitive response to a signal as involving cell memory.[13] Or as C. J. Marshall explains: "It is perhaps surprising that there is a relatively small number of core signaling pathways, so some of them are used for multiple different outputs. This means that cellular context—the developmental or physiological history, together with what other signalling pathways are active—determines signalling output. I think we still have a lot to learn about how cell history and signal integration determine output" (Marshall in Hynes et al., 2013, 395). Joan Massagué refers to this as "context-dependent plasticity" (Massagué in Hynes et al. 2013, 395).

Many of those who use computer and electronic metaphors to think about how cells communicate with one another recognize the disanalogies, but either consider them to be of little concern or unavoidable. One leading biologist in the area openly defends the employment of such metaphors as an important means for humans to achieve understanding of how cell signaling works. "While there is a long way to go in filling out the biological wiring diagram, computers, simulations and metaphors have helped the human mind to keep pace so far" (Bhalla 2003, 63). In fact, the internal dynamics of cells are proving to be so complex, says Bhalla, that they may ultimately be beyond intuitive grasp. However, he explains,

> The process of specifying biological complexity in a reductionist manner leads to parallels with other, better understood systems. Such parallels and metaphors provide several possible paths for thinking about signaling networks. Thinking in metaphors may be a hazardous way of drawing scientific conclusions, but combined with numerical simulations and experimental anchors, it may well be the best way the human mind can usefully grapple with biological complexity (Bhalla 2003, 47).

But others, pursuing what is described as "network-based medicine" guided by a systems biology approach, argue that "the concept of linear cascades provide [sic] a limited and even misleading conceptual frame-work to determine how signal transduction is studied. This in turn un-derlines the importance of a shift in how we conceptualize information processes and a discontinuation in the use of simplistic pathway dia-grams and instead move toward context-dependent and probabilistic concepts" (Jorgensen and Linding, 2010, 19).

5. Proteins and other signaling "machines"

The proteins that are constitutive of many intracellular structures, such as the ribosome, the apoptosome (a complex involved in some forms of mitochondria-mediated programmed cell death), and other func-tional protein-complexes are known to have well-defined quaternary structure, and for this reason are commonly described as protein "ma-chines."[14] In a much-cited article, the biochemist Bruce Alberts argued that current and future biologists should think of the cell as a factory consisting of an "elaborate interlocking network of assembly lines, each of which is composed of a set of large protein machines" (Alberts 1998, 291). The analogy is justified, he argued, because like human-made machines "protein assemblies contain highly coordinated moving parts" within which "intermolecular collisions are not only restricted to a small set of possibilities, but reaction C depends on reaction B, which in turn depends on reaction A—just as it would in a machine of our common experience" (291). The popularity of this comparison means that the protein complexes involved in signal transduction pathways are also commonly thought of as having a similar machine-like nature. However, some biologists have criticized this metaphorical conception of signaling complexes as inadequate and misleading.

Bruce Mayer and his team at the Richard D. Berlin Center for Cell Analysis and Modeling at the University of Connecticut investigate the spatial and temporal organization of molecules in live cells *in vivo*. In a 2009 paper, they write: "The analogy between cell signaling and man-made machines is all-pervasive, frequently adopting the imagery of elaborate clockwork mechanisms or electronic circuit boards . . . But is this really an accurate, or useful, description of the actual processes used by cells?" (Mayer, Blinov, and Loew 2009, 1). They argue it is not, because signaling complexes are most likely to be highly dynamic, ephemeral, and pleiomorphic associations of proteins rather than the highly stable proteins with well-defined quaternary structure originally called "machines." In contrast to the machine hypothesis, they suggest that many signaling complexes, such as membrane receptors, are highly

variable structures that look "less like a machine and more like a pleiomorphic ensemble or probability cloud of an almost infinite number of
possible states, each of which may differ in its biological activity" (2).
This would mean that the output of some signaling complexes might be
a fuzzier signal than a clearly discrete one. It is very unlikely, they argue, that transmembrane receptors, such as that for the platelet-derived
growth factor (PDFG), exist in only two binary states, activated or un-
activated. The receptor has multiple (at least ten) autophosphorylation
sites, which are inactive when there is no ligand or signal bound to the
receptor. But when a ligand is bound, it is unlikely that all sites capable
of being phosphorylated are so at the same time and so there is reason
to believe that any two activated receptors will be phosphorylated at
different sites and in different combinations. Because the receptor for
PDGF dimerizes or combines with other PDGF receptor proteins when
activated (a feature shared by other members of the receptor protein
kinase family), this means that for any activated receptor dimer there
will be an even larger number of possible states. Things get even more
complicated, since phosphatases are constantly removing phosphate
groups from the intracellular domains as quickly as they are being phosphorylated, and which of the one hundred or more cytosolic effector
proteins competing with one another for a binding site will end up binding to the receptor so as to transmit a signal further into the cell depends
on which sites on the receptor are phosphorylated, the local concentration of effector proteins, and their relative affinities for binding at the
available receptor sites. In light of this "combinatorial explosion of possible states" and the fact that many of the protein-protein interactions
involved in the formation of signaling complexes are of modest affinity,
so that signaling complexes will be ephemeral associations that "flicker
rapidly between many different states" (6), many signaling complexes,
they conclude, are distinctly unmachine-like.

A key barrier to greater recognition of the inadequacy of the machine and electronic circuit metaphors that underlie the "typical 'car-
toons' of signaling pathways with their reassuring arrows and limited
numbers of states," they believe, is "the lack of a good analogy from our
daily experience" (Mayer, Blinov, and Loew 2009, 6). Biologists "are
naturally more comfortable thinking of mechanical devices with states
that are clearly defined and limited in number" (7), but this tendency,
they argue, is not up to the serious conceptual challenge of studying and
comprehending a very different sort of biological system.

Part of the problem is that most of the standard techniques used to
investigate signaling pathways involve the unnatural overexpression of

target molecules or provide information on average states of a population of target molecules only, so there are genuine concerns about the results being artifactual. Recent preliminary confirmation for the "pleiomorphic ensemble hypothesis" over the "machine hypothesis" has been provided by means of computational modeling capable of dealing with the combinatorial complexity involved in the mitogen-activated protein kinase (MAPK) pathway of the yeast pheromone signaling network (Suderman and Deeds 2013). The authors of this study conclude that there is likely a spectrum of structural motifs in signal transduction from those very machine-like to others of a highly transitional ensemble nature (10).

Another important kink in the analogy between proteins and machines has only recently become apparent, namely the existence of proteins with no regular stable three-dimensional (i.e., tertiary and quaternary) structure. Intrinsically disordered proteins (IDPs), as they are known, appear naturally in the cell and seem to have significant physiological function (including the formation of signaling complexes) despite their lack of regular structure (Dunker et al., 2001; Tompa 2012). The existence of a structurally disordered yet functional protein presents a challenge to the construal of proteins as standardized machine-parts. This is just one of the reasons cited in a recent review article for dissatisfaction with the current practice of representing what is known of signaling network architecture by means of simplistic visual line drawings (Lewitzky, Simister, and Feller 2012). The depiction of complex protein-protein interaction networks by means of "furball" diagrams and signaling pathways and networks by means of "dumpling soup" diagrams are said to bear "in their simplicity a striking similarity to children's drawings" (2740). The authors suggest that adequate representation of the highly complex and dynamic behavior of signaling components in spatial and temporal dimensions will likely require sophisticated visual methods not unlike the technology used in the creation of 3-D films (2747).

6. Return of agential metaphors: cell signaling as "anarcho-syndicalist" cooperative

Signaling networks are, then, highly dynamic processes that belie the implication of a static and deterministic entity suggested by the pathway and circuit metaphors. For this reason, alternative metaphors have been suggested of a more social or sociological nature highlighting the "cooperativity" of signaling components, in analogy to semiautonomous agents interacting within an intracellular "ecology." In a specific bio-chemical sense, the multiple ligand-binding sites of a protein are said to be cooperative if the binding of a ligand molecule at one site increases (or decreases) the affinity of ligands of the same type to other binding sites (Alberts et al. 2008). But the notion of cooperative behavior in proteins can be construed more broadly to denote the fact that proteins as a whole often come together to form larger signaling complexes. When this occurs, the individual proteins are commonly said to be "recruited" into "playing a role" in these large and dynamic signaling complexes, which after a time may disassemble again into the individual components. So rather than the strictly hierarchical and static logic of an electronic circuit board, one critic of the standard image of linear pathways and circuit boards declares, "There is no dictator in cell regulation, no first among equals, no master regulator, no top-down system of governance. The time has come to acknowledge that the cell is anarcho-syndicalist" (Gibson 2009, 480). The point of describing signaling proteins and other components as cooperative is to highlight their similarities to agents with flexible behavior. Protein kinases, phosphatases, and other signaling molecules have been described as possessing "social abilities" not only because of their tendency to interact but because of their ability "to act in a flexible, unscripted manner—another feature of adaptive agents" (Fisher, Paton, and Matsuno 1999, 164). One could describe the transistors, conductors, and other components in an electric circuit as cooperative, but this would seem odd since they show no movement or flexibility in their behavior. In contrast, one might reasonably compare the behavior of signaling components to the complex set of alliances and counteralliances among a group of people in a medieval court, in which the behavior of an agent is sensitive to the presence and absence of other agents ("The enemy of my enemy is my friend; at least while my primary enemy is around, that is.")[15] This phenomenon of shifting alliances and flexible behavior seems aptly characterized as what some are calling "molecular sociology" (Robinson, Sali, and Baumeister 2007). In fact, Martin Rodbell, the discoverer of the role of G protein-coupled receptors in signal transduction, also thought of cell communication in social terms: "biological communication consists of a complex meshwork of structures in which G-proteins, surface receptors, the extracellular matrix, and the vast cytoskeletal network within cells are joined in a community of effort, for which my life and those of my colleagues is a metaphor" (Rodbell 1995, 221).

If one looks at the titles of research papers in the field of signal trans-duction, one is immediately impressed by the predominance of verbs and action-related terms. Two recent titles from the journal *Science Signaling* will suffice to illustrate:

> "Oxidative DNA Damage Induces the ATM-Mediated Transcriptional Suppression of the Wnt-Inhibitor WIF-1 in Systemic Sclerosis and Fibrosis."[16]

> "Phosphorylation of the TATA-Binding Protein Activates the Spliced Leader Silencing Pathway in *Trypanosoma brucei*."[17]

These titles indicate a complex community network of interacting agents that oxidate, damage, induce, mediate, transcribe, suppress, inhibit, phosphorylate, bind, activate, splice, lead, and silence one another. In order to deal with this entangled system of activity, some systems and computational biologists have even employed models originally developed by social scientists to understand the dynamics of the relationships between striking mill workers and their employer to gain insight into the complex network of signaling crosstalk that goes on in a human cell (Farkas et al., 2011). Similarly, Börlin et al. (2014) employ an "agent-based model," with rules governing the behavior and interaction of individual molecular components (agents) of the cell autophagy pathway, that not only realistically captures short-term dynamic behavior but also predicts novel long-term behavior.

So despite the prevalence of protein-machine talk, it appears that scientists also find it difficult *not* to think of the activities of proteins and other subcellular molecules in analogy with the behavior of human agents.[18] As the philosopher of biology Lenny Moss observes, "molecular biology is now beginning to reveal the extent to which macromolecules, with their surprisingly flexible and adaptive complex behavior, turn out to be more *life-like* than we had previously imagined" (Moss 2012, 170). Although this stands in contrast with the discourse of protein machines, it is interesting to note that it is common for researchers to speak of signaling circuits, wiring diagrams, etc., and of cells "making decisions" in the very same paragraph.[19] In that sense, it is a bit inaccurate to speak of a "return" of agential metaphors, for our image of the cell has for over a century now been a hybrid of machine and agential metaphors. Some philosophers worry about this anthropomorphizing of molecules. Peter Godfrey-Smith (2009), for instance, is critical of the special type of "agential narrative" lying behind the gene-centric view of evolution and of biology more generally that is best illustrated by talk of "selfish" genes and DNA. Such talk can lead to a Darwinian paranoia, where we humans begin to worry whether we are really in control or whether we are merely instruments being used by "a hidden collection of agents pursuing agendas that cross-cut or oppose our interests" (Godfrey-Smith 2009, 144). Similarly, Dan Nicholson (2014b) cautions against what he describes as a "molecular animism" that relegates causal agency properly belonging to the entire cellular system to a privileged few molecules that are given special credit when described as "regulators," "integrators," "organizers," and so forth.

The recent emphasis of the flexible behavior of biomolecules has, Moss (2012) argues, implications for philosophical accounts of mechanism and explanation, not to mention biomedical research into drug development and therapy.[20] For if this newer way of thinking about intracellular signaling is correct, then these molecular mechanisms are decidedly un-machine like, and philosophers of mechanistic explanation may need to rethink what they understand the concept of mechanism to be.[21]

On the other hand, it has become apparent in the last couple of decades just how much internal organization there is in signaling pathways and how significant the compartmentalization of key signaling molecules to specific locations in the cell can be for the function and results of intracellular signaling. Scientists are once again emphasizing that the cell is not a loose bag of signaling molecules and target substrates that interact by random diffusion. While the cell interior is in some respects less a "thin watery bouillon soup as a thick fish chowder" (Bray 2009, 92), it is also in other respects a highly structured environment, both spatially and temporally. Many of the components of signal transduction pathways are held together in close spatial proximity by special "scaffold," "anchor," "hub," and "adaptor" proteins, which by selectively binding other proteins or lipids at so-called interaction domains (sequences of typically forty to one hundred amino acids) increase the likelihood and efficiency of various signal components interacting so as to enhance the speed and specificity of a signal by either avoiding or facilitating crosstalk with other pathways.[22] Furthermore, the cell cytoskeleton and the extracellular matrix (ECM)—a network of proteins and polysaccharides secreted by cells that provides mechanical and biochemical support to cell and tissue organization—both help to convey signals, acting as a nearly literal communications network (Forgacs 1995). So if some researchers are critical of the common depiction of signal networks as static hard-wired electronic circuits, others object that they are presented as too loosely structured (think "dumpling soup"). For instance, Stephan Feller, a molecular oncologist and editor in chief for the journal *Cell Communication and Signaling*, complains that,

Even in the latest editions of top cell biology books the cell signalling machinery is typically depicted as an assembly of fairly
unorganised protein molecules, for example diffusing more or
less freely in the cytosol. According to current textbook wis-
dom, upon activation of a signalling pathway its components
stochastically meet to generate transient assemblies in the form
of signalling "cascades" or protein complexes with up to 10 or
so components. These in turn appear to be linked together into a
giant "floating signalling network" of several thousand proteins
which nobody really understands (Feller 2010, 1).

Feller suggests that the molecular components of signaling complexes are concentrated differentially in specific pockets of the cell or to
the cell membrane so as to increase the likelihood of their interacting,
thereby overcoming the improbability of their meeting by random diffusion alone (making the point once again that the cell is not a bag of
chemicals). While new technologies like high-throughput proteomics,
transcriptomics, and genomics hold great promise for future advances, Feller emphasizes that all this data must be organized through
some conceptual framework or other. He predicts exciting times ahead
for young researchers, on two conditions: (1) that they leave behind
the older dogmas and misconceptions that have arisen from the use of
"primitive" tools of analysis, and (2) that they resist "the apparent urge
of the human brain to build simple linear models with a small number
of components to explain functional relationships" (Feller 2010, 2).

The importance of spatial organization within the cell by means of
scaffolding, anchor, and adaptor proteins, however, suggests the analogy between electronic circuits and intracellular signaling may not be
entirely inaccurate—although it is to be noted that biologists also frequently talk of interaction domains as promoting social behavior and
"teamwork" among proteins.[23] But an even stronger case for the analogy between cells and electronic computers is made by the successful
construction of artificial "logic gates," gene regulatory switches and
other devices in bacterial cells by so-called "synthetic biologists."

7. Engineering metaphors and synthetic biology

Synthetic biology may be succinctly defined as "(A) the design and construction of new biological parts, devices, and systems, and (B) the redesign of existing, natural biological systems for useful purposes."[24] As
the fundamental unit of life, much of this work is performed at the level

of the cell, either as the "locus of inquiry" or the "subject of inquiry," to use Bechtel's helpful distinction. Characteristic of those doing synthetic biology is an engineering approach to living systems. The aim is not just (or even primarily) to understand how living cells work in their natural states and environments but to make of them what Jacques Loeb had dreamed in the early years of the twentieth century: a "technology of living substance" (Landecker 2007).

In 2000 a team of researchers at Boston University announced in the journal *Nature* that they had successfully constructed the genetic equivalent of an electrical toggle switch in a bacterium (*E. coli*) that could turn on and off genes upon the reception of the proper chemical or thermal signals. As the authors explained, this was only the first of many artificial devices they hoped to create: "Inspired by electrical circuits as well as natural biomolecular networks, these devices include timers, counters, clocks, logic processors, pattern detectors, and intercellular communication modules" (Gardner, Cantor, and Collins 2000, 1248).

One of the creators of the genetic toggle switch recently reflected on the real significance of the invention.

> The beauty of the toggle switch is that it gives cells a memory. Before the toggle switch, if scientists wanted a cell to switch a gene from on to off or vice versa, they would have to continuously give it an inducer for the gene encoding that protein. This is like having to hold your finger on a light switch to keep it on, which is not very useful if you want to move around the room. The toggle switch, however, keeps a gene switched on with one single delivery of an inducer. It gives the cell a memory of the state it should be in. For companies that need inducers to turn on the production of a protein inside cells, this method means it can spend less money on inducers (Collins 2012, S9).

The suggestion that these novel cell-constructs are designed with application by private companies in mind is significant. Synthetic biology is in many ways the latest stage in development of the biotech industry, an industry that has so far failed to produce the much-anticipated successes promised through the insertion and deletion of individual gene-segments alone. In light of growing recognition of the link between miscommunication in signaling pathways/networks and many human diseases, much research into signal transduction by synthetic biologists is now carried out with an eye toward biomedical application.

Substantial progress began a little over a decade ago with the creation of synthetic gene networks inspired by electrical engineering. Since then, the field has designed and built increasingly complex circuits and constructs and begun to use these systems in a variety of settings, including the clinic (Ruder, Lu, and Collins 2011, 1248).

Much of the biomedical research conducted in the spirit of synthetic biology is articulated in terms of "rewiring cellular circuits" and manipulating gene-regulatory switches (e.g., Bashor *et al.* 2010). Synthetic biology is an interventionist scientific project, and not so much about pure representation or "mirroring" nature as it is about remaking it. It is, therefore, a characteristic example of what is called technoscience. O'Malley et al. (2007) and Keller (2009a) have noted the epistemic assumption driving much of synthetic biology is that "making is knowing," a thesis frequently expressed with a popular slogan attributed to the theoretical physicist Richard Feynman (1918–988): "What I cannot create, I do not understand." There is an understandable appeal to the idea that if we can recreate a phenomenon under controlled conditions, then we understand precisely how and why it occurs. The quotation from Feynman, however, goes further, suggesting that the ability to recreate something at will is not only a sufficient condition for understanding it but also a necessary one.[25] However, biology has been famously more difficult than at least some areas of physics because many of its phenomena are multiply realizable, meaning that there is more than one way for a phenomenon to come about. This has been an important lesson learned by molecular geneticists in the last few decades. For example, while it is convenient to talk of "the gene for . . . ," it is now understood that in the majority of cases the map between genotype and phenotype is not one-one, but many-one (polygenic) or one-many (pleiotropic); throw in the extra complexity of alternative splicing, which means that from one and the same DNA sequence different protein-products can be synthesized, depending on how the RNA transcripts are "spliced" together to form a "readable" transcription sequence, which is translated by the ribosomes into a three-dimensional polypeptide sequence, and is then subject to further "post-translational" modifications. The upshot is that, just as we saw in chapter 2 in the case of biochemists' efforts to understand *in vivo* cell metabolism by means of investigation of *in vitro* cell-free systems, the ability to synthesize a phenomenon under highly controlled and artificial conditions—for instance a synthetic genome or an artificial cell (e.g., Karzbrun et al. 2014)—does not necessarily equate to an adequate understanding of what is going on in the highly complex and messy environment of the living cell and living organism situated in an equally complex and dynamic external environment.

Because synthetic biology is principally concerned with designing and building—activities intimately associated with the field of engineering—it is not surprising that its preferred metaphors for cells and subcellular components include terms like circuits, switches, wiring diagrams, and so forth. These *technomorphic* metaphors (as opposed to being anthropomorphic) are particularly suited for synthetic biology's quest to reengineer cells for specific human purposes: medical, industrial, and commercial.

This points again to a less obvious aspect of metaphor use. As we saw above with the example of the factory metaphor, metaphors are not only descriptive tools. Metaphors are also prescriptive: they encourage and facilitate particular approaches to the object under study. The engineering metaphors suggest a project, of rewiring the cell's signaling circuits. If we use electronic and computer metaphors to talk and think about cells, it should not be surprising that we try to reengineer, redesign, and rewire them. Synthetic biology and its project of creating potentially patentable commodities from living cells and biomolecules are facilitated by the description of cells as electronic gadgets and devices. If we choose to think of cells as just another form of device, then the licensing of patents or copyrights for their manipulation and "improvement" is a natural consequence.[26]

Intellectual property law scholar Graham Dutfield (2012, 173) notes that the fact that so many creative achievements in the life sciences have been patentable "has much to do with how they are described" and that engineering metaphors play a significant role in this trajectory. The talk of rewiring and reprogramming cells and signaling circuits no doubt are examples of what Nelkin (1994) calls promotional metaphors: they serve to encourage optimism about the likelihood of success in the project among other scientists, government regulatory bodies, and potential financial investors. However, even advocates of the synthetic biology industry have reservations about the use of engineering metaphors to describe living cells and organisms. While discussing the question whether synthetic biology should be externally regulated or guided by its own set of voluntary codes, members of the Biotechnology Industry Organization concede that:

Metaphors utilized for synthetic biology have often been based on electronic toolkits—i.e., systems that are modular and open to reconfiguration. However, these metaphors can mislead public perception of biotechnology because living organisms are not directly analogous to modular electronics, and therefore, law, policy, and research and development in synthetic biology should not be modeled after law, policy, and research and development in the fields of computer science and electronics (Erickson, Singh, and Winters 2011, 1254).

As useful as those metaphors may be for the sort of interventionist and commercial ventures pursued by profit-seeking biotech companies, they may not be as adequate for other projects, for instance, understanding how cells and organisms naturally develop into differentiated cells and organisms. It is for this reason that Keller worries that the fruits of synthetic biology may be "understanding no, changing yes" (Keller 2009b, 300). And because many of its practitioners operate by reducing biological complexity so as to create more manageable and reliable "standard parts," O'Malley et al. (2007, 61) express concern about what they call a "central tension in synthetic biology, between construction and comprehension."

Metaphors are sometimes referred to as cognitive instruments, and like any instrument they may introduce artifacts into our understanding of an object or system. They can lead us astray by encouraging us to see those aspects of the target system that are most similar to the metaphor's source domain, but to be blind to those aspects that do not fit the metaphor. And as we have seen above, metaphors can change not only the way we think about or see things, they can lead to real material change in the thing in question by encouraging us to reshape, rebuild, or rewire it to better fit the image promoted by the metaphor.

Although it seems not to be well recognized by scientists and science journalists, the decision to use a particular metaphor is a prescriptive speech-act. To speak of cell circuits is not just to offer a description of cells, it is (at least implicitly) to suggest that cells are devices that can be—and perhaps even *ought* to be—reengineered to suit our needs or desires. But there are dangers inherent in a rush to intervene in mechanisms of cell signaling without understanding the broader organismal and ecological contexts within which cell communication plays out. The practice of describing cells and their communicative capabilities in the engineering language of computer electronics lends itself nicely to attempts by the pharmaceutical industry to isolate and exploit for commercial gain elements of specific signaling pathways (typically ligands or receptors) as though they were not integrated elements of a dynamical network within a living organism, where crosstalk is proving to be the norm. Here too, the engineering metaphors are likely playing the role of promoting the project to potential investors and regulators. But the riskiness of this practice is illustrated by the recent case of Merck's "miracle" drug Vioxx.[27] In this case, the drug (rofecoxib) was intended to inhibit the enzyme COX-2 (cyclooxygenase 2) implicated in the body's inflammation response associated with the pain of arthritis. However, it also had the side effect of increasing the risk of heart attack and other vascular events, because in blood vessel cells it crosstalked with another pathway downstream from COX-2 involving prostacyclin, a signaling molecule involved in the inhibition of blood platelet aggregation. Anyone who has read a magazine advertisement or watched a television commercial recently for a drug has likely been struck by the long list of counterindications and possible side effects about which consumers must be warned. In light of the existence of crosstalk between signal pathways and the highly complex, dynamic, and context-sensitive (e.g., post-translational modifications of proteins and epigenetic influences) nature of cell signaling, should we really expect to be able to tinker with the functioning of cells and organs in a predictable and safe fashion, the way we do with machines and electronic computers?

It is not my intention to censure or even to discourage the use of computer and engineering metaphors in this or any other branch of biology. One promising means of designing safer and more effective drugs involves the development of quantitative mathematical and computer models of signal transduction networks.[28] But might efforts to create models of the cell *in silico* only further entrench the computer circuit metaphors? One of the motives driving the rise of computational and systems biology is the recognition that cell physiology is more complicated than the simple circuit pathway diagrams can represent, and that they offer little in the way of predictive power. If nothing else, the trend to emphasize that all signal pathways are parts of larger and more complicated networks shows that scientific research can be assisted by the adoption of more adequate metaphors. As I will argue in chapter 6, science has always benefited from a division of labor among researchers and the encouragement of multiple paths of inquiry guided by different metaphors. Scientific understanding is furthered as much by failure as by success, and pushing a metaphor to its limits is part of the scientific process that the philosopher Karl Popper called "conjecture and refutation."

But where failure or the possibility of being misled has ethical implications, we need to be more careful in our choice of metaphors and how
strongly we allow ourselves to be guided by them. What the bioethicist
Cor van der Weele calls an "ethics of attention" requires us to be cog-
nizant of the moral and social implications of the choices of explanation we make for biological and biomedical phenomena (Van der Weele,
1999, 133ff.). It is in this respect that the implicitly prescriptive nature
of metaphor use needs to be more widely recognized and discussed.[29]

Conclusion

We began this chapter with E. B. Wilson's declaration at the beginning
of the twentieth century that the biological question of greatest significance concerned "the means by which the individual cell-activities are
coördinated, and the organic unity of the body maintained" (Wilson
1896; 1900). We have seen that the answer provided over the span of
the last century invokes the idea that the cells of the body communicate with one another. Through intercellular communication, cells of
the developing embryo nudge and coerce one another into different developmental pathways leading to different cell fates as distinct types of
tissue-cells, which are arranged into distinct organs and organ systems
whose collective activities are mutually controlled and coordinated for
the benefit of the collective organism as a whole. In this way, the society
of cells exerts its social control (chapter 3), coordinating and regulating
the various activities of its individual members in a way analogous to
that of a bustling and complex human city or nation.

The details of this story of intercellular communication and coordination by means of chemical hormones, electrical signals, and physical
nudges, twists, and stretches took us down a level deeper to the molecular account of intracellular events at which various proteins play a vital
role. The key metaphors with which this scientific account of cellular activity is articulated fall into two main types: machine/computer metaphors and agential/social metaphors. Of the first type, we discussed:
*signal, receptor, transduction, circuit, programming/reprogramming,
wiring/rewiring, switches*, protein *machines, adaptor* proteins, *anchor*
proteins, *scaffold* proteins. Of the second type, some of the metaphors
we discussed included: *messengers, cooperativity, strategies, surveillance, teamwork*, and an assortment of action verbs attributed to proteins and other cell components such as *inhibit, regulate, recruit, evade*.
Several of the key metaphors at the heart of the science of intracellular
signaling do not quite fit either category but are closely associated with

the computer engineering approach. These were the metaphors *cascade*, *pathway*, and *network*. The network metaphor, which is on the ascent in both the systems and synthetic biology approaches, can also have a social connotation.

The two classes of metaphor seem to play distinct roles within the scientific discourses considered here. For instance, when scientists are talking about *what* proteins do, they tend to use action verbs and agent metaphors; when they talk about *how* they do it, they frequently use machine and engineering language. Although one might expect to find a natural affiliation for engineering and machine metaphors among synthetic biologists—who are quite explicitly interested in manipulating or reprogramming cell behavior, use of the two classes of metaphors does not follow a simple interventionist vs. "pure" science division. While synthetic biologists do tend to use lots of engineering metaphors, they're not the only ones doing so. Experimental intervention into a system is now a standard and typically effective means of gaining greater understanding of how the system naturally operates, and engineering-machine metaphors help facilitate this approach.

Biologists from all disciplines and approaches show a disposition to talk of cells (and of their components in particular) as though they were a curious hybrid of agents and machines. The example of the children's cartoon-turned-Hollywood movie *Transformers* comes to mind—proteins are described as if they were tiny Transformers, robot-agents that assemble and disassemble to create larger machines to accomplish difficult tasks that can only be achieved by collective and cooperative teamwork.

Metaphors from the two main classes seem to be used by scientists to highlight different aspects of the same general phenomenon, so that they are not necessarily being used to offer competing accounts but rather complementary descriptions. For even though mechanistic and computer engineering metaphors have come to provide a central organizing background or framework for how scientists think about cell-cell communication, there is still a mix of metaphors in vogue as scientists continue, to give just one example, to talk at once of both signals and messengers.

There has been a movement in philosophy of science in the last few decades away from *a priori* assumptions about the "unity of science" toward a more "naturalistic" attention to the actual diversity of practices in the various sciences. This has been accompanied by a greater emphasis on the role of models in scientific inquiry and explanation. Rather than the older picture of intertheoretic reduction of the vocabulary and explanations of one theory to the vocabulary and principles of a more fundamental and "deeper" theory, philosophers nowadays note how scientists (especially in the biological and biomedical fields) frequently employ a plurality of models and techniques with distinct vocabularies and assumptions to piece together an overlapping patchwork of explanations and understanding of the world (Cartwright 1999; Blasimme, Maugeri, and Germain 2013; Baetu 2014; Green 2013).

The picture of science emerging from this greater attention to the diversity of models, techniques, and explanatory frameworks is not that of one theory or model or paradigm or vocabulary achieving domination over all its competition, but one according to which science invokes multiple views of a phenomenon as the situation demands. Rather than working with only one theory or paradigm or metaphorical vocabulary at a time, scientists often proceed by hook *and* by crook. Scientists are not, after all, trying to describe a homogenous sphere for which one account in one vocabulary might suffice—they're trying to understand in the cell an extremely complicated system of very complicated bits with a breathtaking multitude of aspects. And of course, they are not only attempting to describe it, it's not a "just look, don't touch" situation; some are also attempting to fix it when it's not working properly, while others attempt to alter and improve it for a whole variety of purposes. These issues will be taken up at greater length in chapter 6.

Metaphors, it is often said, are of value to scientists because they offer a novel view or perspective from which to see the thing in question. Alternatively, metaphors are described as tools or instruments scientists use to investigate the world and with which they devise explanations and achieve greater understanding. Both of these standard accounts of why metaphors are useful in science are themselves reliant on metaphor. Do these particular metaphors about metaphor—what I will call *meta-metaphors*—help us to understand how metaphors work in science and why they may be valuable or misleading? Might these meta-metaphors themselves be more misleading than informative? Need we resort to metaphor at all to provide an account of how metaphors work in science? I turn to these questions in the next chapter.

5   Metaphors in Science: "Perspectives," "tools," and other meta-metaphors

> A picture held us captive. And we could not get outside it, for it lay in our language and language seemed to repeat it to us inexorably.
>
> **Ludwig Wittgenstein,** *Philosophical Investigations* ¶ 115

1. Introduction

I hope to have sufficiently demonstrated that metaphor has played a significant part in both our past and current understanding of cell biology. There are plenty of other metaphors that I haven't discussed, partly because others have already done so.[1] Other areas of science, such as physics, chemistry, geology, and noncellular biology, also reveal widespread use of metaphorical language and thinking. Science in general is full of metaphor. But are the metaphors essential to the science? Perhaps, some will insist, scientists resort to metaphor only as a matter of convenience, for the sake of expressing unfamiliar ideas to the nonprofessional public in terms which are more familiar to it, or as a kind of dumbing down of difficult content. While this is indisputably true of some instances of scientists' use of metaphorical language, its widespread occurrence in the professional literature, where scientists are writing for one another and for themselves, shows that there is more to it than that. Those who would defend the thesis that science aims to provide a uniquely true and canonical description of reality may concede that metaphor is an efficient facilitator of analogical reasoning, and in that case may be important for the creation and development of new ideas or hypotheses; however, they may insist, its role is restricted to the context of discovery and has no real cognitive function in the context of justification, where the critical and logical evaluation of argument, assessment of evidence, and eventual formulation of explanations occur. On this view, metaphor is—or ought to be—wholly eliminable from scientific discourse. Scientists may continue to speak in metaphorical terms in review articles and even in reports of original experiments and research, but this is merely a convenient shorthand, since one could always replace each instance of metaphorical language with an equivalent "literally true" alternative or an explicitly clear simile at the least. It is difficult to assess such a claim about what could be done in principle, if not as a matter of regular practice. Because it concerns what science could be like in the ideal, it is of little relevance for attempts to understand how science actually does get done; and in the spirit of the naturalistic turn in philosophy of science (and in science studies more generally), I prefer to study how scientists actually go about creating knowledge, theory, and explanations.[2] What the evidence shows is that scientists working in the area of cellular biology think about the systems they study using a diverse range of metaphors, and for a diverse range of projects: some primarily concerned with representing cells and organisms, others more obviously aimed at intervening in and reconstructing them.

Three basic roles have been claimed for metaphors to play in science: (1) a rhetorical or communicative role (which would include pedagogical purposes and talking to nonscientists); (2) a heuristic function in the creation of new ideas and hypotheses; and (3) a cognitive or theoretical function in the formulation of scientific explanations (e.g., Bradie 1999). I wish to add a fourth: (4) a technological role as cognitive tools of intervention and manipulation. This fourth role, it seems to me, has been underappreciated in the scientific metaphor literature.

Clearly, any of the instances of metaphor we have looked at can and do fill the first function. That they have also played a significant part in the creation of new hypotheses and ways of thinking about cells and organisms and how they work I trust has been sufficiently demonstrated in the previous chapters. The real point of contention for those interested in whether metaphor is an essential element of science or not concerns the purported third function. Can metaphors play a legitimate role in the critical evaluation of evidence and theory, and in the development of scientific explanation? Can metaphors result in proper knowledge and understanding? I argue that they can, while at the same time providing a critical assessment of some of the more popular accounts of why metaphors are of value to science, the majority of which themselves resort to metaphor in their explication, and thereby raise questions about their general adequacy.

The two most common accounts in this line are (1) that metaphors provide a novel or useful perspective from which to see a subject (or alternatively a lens or filter through which we see an object), and (2) that they function as cognitive or conceptual tools. While lenses and filters qualify as types of tool, the perspective account aligns well with the traditional empiricist association of knowing with seeing and the thesis that understanding involves a kind of mental vision or sight that results in an internal picture of an external world. The tool account is more congenial to the pragmatist thesis that at least some (and perhaps all) knowing is ultimately about being able to *do* things successfully, to intervene from *within* the world and to manipulate objects and tools in order to achieve goals we humans find worth pursuing. I will refer to these two fundamental metaphors—or meta-metaphors—as the *perspective* and the *tool* accounts of metaphor.

2. How do metaphors work?

Modern philosophical discussion of metaphor is often traced back to the work of Max Black (1909–88), who was himself drawing on the earlier ideas of I. A. Richards (1893–1979). Black's (1962a) "interaction" theory of metaphor took issue with the thesis that metaphors are merely alternative ways of stating a simile. According to this substitution theory of metaphor (which Black attributed to the logical positivists), to say that "man is a wolf" is just another way of saying that "man is like a wolf—in the following ways: he can prey upon other humans, he can be conniving and sly, etc." On this view, the metaphor merely expresses a set of features with which the two subjects in question are compared and said to share in common. Black argued that metaphors often do more than simply give expression to a set of similarities between the source and target domains[3]; the metaphor, in some cases, actually *creates* the similarities. So it is not that we had already recognized that humans and wolves are alike with respect to features X, Y, and Z, and the metaphor is just a handy way of expressing this. Rather the metaphor encourages us to see certain similarities between the two we hadn't considered before. Moreover, the metaphor does not just draw a comparison between the source and target domains by extending unidirectionally features of the one to the other, it "interacts" with them both, making humans more wolf-like, and at the same time, wolves more human-like.[4] And, importantly, the set of similarities that the metaphor leads us to see between the two is open-ended, the similarities not preexisting the metaphor but getting articulated as the metaphor is used and developed over time.

Because metaphors are not simply shorthand equivalents for similes or other comparisons, Black argued that metaphorical language cannot be so easily expunged from scientific theory and reasoning. There is no literal equivalent for which they can be substituted without loss of content, or more appropriately, without loss of effect. Others have endorsed this claim and expanded upon it, including Mary Hesse; George Lakoff and Mark Johnson; Michael Ruse; Michael Bradie; and Theodore Brown. Mary Hesse (1966) argued that explanations involving the introduction of novel theoretical language are "metaphorical redescriptions," whereby some phenomenon (the explanandum) is redescribed in the terms of a new theoretical language (the explanans). Following Black's claim that "every metaphor is the tip of a submerged model" (Black 1979, 30),[5] Hesse suggested the introduction of a metaphor is the beginning of the formulation of a model. The metaphor "light is a wave," for instance, was the means to the creation of an explanatory model, despite the fact that ultimately a material ether to carry such a wave was deemed unnecessary. The practice of thinking about light as a wave was informed by an analogy with other better-understood phenomena such as sound waves moving through a gaseous medium and waves of water on the ocean (Hesse 1966). Not all scientific explanations involve metaphor, according to Hesse, but ones that invoke new theoretical language do (Hesse 1966, 161–62). Kuhn (1993) and Boyd (1993) have also argued that the introduction of new theoretical language involves the invocation of metaphors as a means of "nondefinitional reference fixing" in situations where scientists have as yet only imperfect epistemic access to the subject of their investigations. Boyd argues that such metaphors can even be "theory-constitutive" in the sense that they become integral to how scientists carry out and understand their research by helping to identify what its key terms denote. It seems quite plausible to consider Schleiden's and Schwann's use of the term cell (or *Zelle*) as serving this function; through it they encouraged others to look through their microscopes for similar structures and to interpret the development, anatomy, and physiology of plants and animals in its terms.

Michael Bradie (1998) has argued for the extension of Hesse's account to scientific explanation in general, no matter whether the explanans involves a new theoretical language or not. The significance of this, if it is correct, is that metaphor would then be playing more than just an important heuristic role, it would also have an important "cognitive" function. Metaphor would not be confined to the context of discovery, it would also be at work in the context of justification, if we accept that explanation is a cognitive activity, not just a communicative one. But can a metaphor be explanatory? Recall the remark of the French endocrinologist Claude Kordon that "no metaphor is really explanatory," that it only reflects "the cultural references through which we have been conditioned to decipher reality" (Kordon 1993, 96). Metaphors, he seems to suggest, deal in appearances, not reality. I will take up this question below, but first we must consider the popular accounts of why metaphors are of such value to science.

According to Lakoff and Johnson's (2003[1980]) theory of conceptual metaphor, we find it quite helpful to frame unfamiliar and highly abstract concepts in terms more familiar to us drawn from our experiences in the social and natural world. Lakoff and Johnson show how most of our abstract concepts—those that concern things we cannot directly experience through our senses—are rooted metaphorically in our more familiar experiences as embodied beings in the world. For instance, because we are bipedal animals that normally associate good health with being upright and sickness with being prostrate, we commonly use metaphors of "up" and "down" to describe other more abstract situations. For instance, we say "Things are looking up" to express a positive forecast, or "The economy is in a slump" to indicate that commercial activity is not in a "healthy" state. Their observation that we use more familiar experiences to organize our understanding of more abstract and complex subjects is also borne out by the standard accounts of what it is that metaphor contributes to science. As mentioned above, one of the most popular accounts of metaphor's value to science is that it provides a useful perspective on something or allows us to see the target subject in a novel or more familiar light, or through a lens that allows us to recognize certain features that are not as noticeable without the metaphor. Metaphor, it seems, is itself one of those abstract subjects for which we naturally turn to metaphor to help us understand. Let's take a closer "look" at this account and some of its implications for our understanding of how metaphors in science operate.

3. The perspective meta-metaphor

The perspective account has a very ancient lineage, going back to Aristotle, who in the *Rhetoric* (1410 b33) said that a good metaphor ought to "set the scene before our eyes." This account has been a part of discussions about metaphor by many more modern writers, e.g., Black (1962a; 1962b), Van Steenburgh (1965), Hesse (1966), Turbayne (1970), Davidson (1984), Stepan (1986), Kittay (1987), Paton (1992), Maasen, Mendelsohn, and Weingart (1995, 2), Sismondo (1996), Bradie (1998; 1999), Ruse (2000), Runke (2005; 2008), and Camp (2006), to mention only a few.

According to Max Black, metaphor involves the use of language to bring two separate cognitive domains together in such a way that one is used "as a lens for seeing the other," the metaphorical expression enabling "us to see a new subject matter in a new way [sic]" (Black 1962b, 236). Two more recent instances of this tendency will serve to illustrate its continued presence:

> "Metaphors provoke and give birth to new images; by establishing and reinforcing connections, they encourage us to see in new ways" (Otis 2001, 12).

> "The attraction of metaphors consists in their power to make things and complex facts visible" (Brandt 2005, 637).

Many of the writers who adopt the perspective account seem implicitly to recognize that this is a metaphorical way of describing what it is that metaphors do (they often, but not always, put scare quotes around the key words "seeing" and "perspective," etc.), but the implications of adopting this metaphorical account of metaphor have not been explicitly considered.

The philosopher Eva Feder Kittay, however, has been quite explicit in her adoption of the perspective metaphor as a label for her own account of metaphor:

> I prefer to call the account a *perspectival theory*. To call our theory perspectival is to name it for the function metaphor serves: to provide a perspective from which to gain an understanding of that which is metaphorically portrayed. This is a distinctively cognitive role. Since *perspectival* implies a subject who observes from a stance, we can say that metaphor provides

the linguistic realization for the cognitive activity by which a
language speaker makes use of one linguistically articulated domain to gain an understanding of another experiential or conceptual domain, and similarly, by which a hearer grasps such an
understanding[6] (Kittay 1987, 13–14).

Metaphors work, according to Kittay, by transferring the relations between terms or concepts of one semantic field to another content domain, thereby giving it a borrowed structure if it is presently without
its own, or reorganizing it if it already has one. In this way metaphors
"organize" our thoughts just in the way that viewing an object from
a particular perspective or vantage point imposes an organization or
arrangement of properties on what is seen in our spatial and temporal
experience. With the use of a particular metaphor, certain features are
"highlighted" or "brought to the foreground," while others are "pushed
into the background" or out of "view" entirely.

Metaphor consists, then, in more than just the borrowing of a word
or words: it is the *conceptual organization* of a target domain through
application to it of a structure that is already in place among terms of
another semantic field. In saying that the cell is a factory (recall chapter 2), for instance, we suggest that a whole set of relations among members of the semantic field of factories (inputs, outputs, assembly lines,
shipping and receiving, energy and waste byproducts, etc.) are applicable
to the conceptual domain of cells. The structure of these borrowed relations becomes the basis for analogical inferences and hypotheses about
the target subject. This is an account also developed somewhat earlier
by the cognitive psychologist Dedre Gentner, who considers fruitful instances of scientific metaphors to be those that promote rich relational
or structural analogies between two knowledge domains (Gentner 1982;
1983; Gentner and Jeziorski 1993).[7] Metaphors that merely involve
transferring one or two superficial attributes of one object to another
seldom make for useful analogical reasoning. Hooke's original remark
that the structure of cork plant tissue appeared to consist of small cells
or pores was really rather limited in its inferential implications, whereas
later suggestions that cells are factories, that they contain programs and
signaling circuits, that plants and animals are vast cell-states wherein
the physiological labor of life is divided among specialized cell types,
and that some members of these cell societies commit suicide for the
good of the whole— these metaphors evoked rich systems of relations
between subcomponents of the respective semantic fields. And none of
these suggestive signposts would have been likely if instead Hooke had

coined a brand-new term (I suggested "jex," for example, in chapter 1) and all later investigators had followed his lead by employing literal but less fertile and wholly undescriptive labels.

So, according to the perspective account, metaphors are useful to the extent that they encourage us to see or understand one thing in terms of a set of relations associated with another distinct type of thing. This account permits us to understand how metaphors do what they do without positing the existence of any semantic meaning specially associated with a metaphor. The philosopher Donald Davidson, for instance, argued that there is no metaphorical meaning separate or distinct from the literal meaning of the words and phrases traded on by a metaphor. The creative element of metaphor, he suggested, is not in making a novel meaning, it consists rather in a nonpropositional "seeing as": "Metaphor makes us see one thing as another by making some literal statement that inspires or prompts the insight" (Davidson 1984, 263). If we say "Man is a wolf," we mean literally just this, but we recognize that strictly speaking this is false. What the metaphor does, Davidson argued, is to make us see a human as a wolf. We are encouraged by this literally false statement to begin selecting wolf-like features that might apply to a human. Metaphor achieves what it does, in short, by leading us to see one thing as another.

Could this provide support for the claim that metaphor not only plays an important role in scientific theory and explanation, but that it is actually indispensable from science? Jennifer Runke argues that if "part of a metaphor's effect is better understood as a kind of perception," such that it provides a novel perspective from which we view an object, leading to possibly interesting and empirically fruitful descriptions and experimental manipulations, then this aspect or effect of the metaphor could not be replaced by a literal equivalent statement. She writes:

> By construing some of a metaphor's effect as being perceptual, we can now delimit the part which can be literally explicated from that which can't. We can explicate the relevant ways two objects are similar; so we can capture the conceptual content of the metaphor. However, our explication cannot capture the insight that we experience when our perspective of the phenomena has changed (Runke 2005, 8).[8]

That aspect of the metaphor that cannot be captured conceptually (or propositionally), according to this proposal, would be the *phenomeno-*

logical experience of understanding or conceiving an object in a novel way, in addition to the cognitive effects it has on how we then conceive and understand the object in question. It should be noted, however, that if we accept this account it speaks only to metaphor's role as a heuristic factor in the scientific context of discovery, and does not yet show that metaphors are an essential and irreplaceable element in the context of justification. For that we will need to establish that metaphors serve a fruitful and legitimate role in the achievement of some more recognizably cognitive activity, such as providing explanations and understanding of natural phenomena.

Ruse (2000; 2005) has argued that certain key metaphors in evolutionary theory ("design," "struggle," "selection," "division of labour," "shifting balance") are absolutely indispensable in the sense that without them scientists would never have asked the questions they did nor arrived at the answers and theories they have. But isn't that merely to say the metaphors are heuristically indispensable, that they are like the scaffolding used to erect tall buildings and so can be removed once the solid permanent structure is in place? Ruse admits that perhaps the metaphors could be eliminated, but not without "ripping out a huge amount of what actually physically exists in the theory at any given moment," and no scientist in his or her right mind would wish so to geld their theory (Ruse 2000, 606). Metaphors provide the predictive fertility sought for in a good theory, and further progress would very likely come to a halt without them. As Runke phrases it:

> Metaphors allow us to ask some questions that otherwise could not be asked in literal language because in order to ask those questions we need to adopt a particular perspective. Only from a particular perspective will some features seem salient and only with the metaphor will we have the appropriate language that is needed to talk and think about those features (Runke 2008, 190).

If it is correct that metaphor works by the transference of relations or an ordering of concepts and terms from one domain to another, then as Kittay has argued, "to the extent that the speaker has no other linguistic resources to achieve these ends, the metaphor is irreplaceable" (Kittay 1987, 301). And as Elisabeth Camp (2006, 18–19) argues, even if we can eventually replace a metaphor with a literal description, without the original metaphor we may not have a concept at all to replace with the literal term if the metaphor has played an essential role in creating the concept or thought in question. In that sense, that we can replace the metaphor with a literal description or term does not show that the metaphor is eliminable in this generative sense.

Grant (2010) surveys this and several other arguments for the supposed indispensability of metaphor. He argues quite persuasively that these arguments confuse the necessity of our having a concept that is applied metaphorically with the conclusion that the metaphor itself is necessary in order to express a certain thought. For instance, if we talk about one protein "recruiting" another so as to form a component of some larger signaling pathway, we are relying on the metaphor that proteins are humans or social agents.[9] And surely, we could not talk or think about proteins in this way without the concept of a human or a social agent and how they enlist or recruit one another's aid in achieving goals. But the metaphor does not create the concepts of a human, a social agent, or of recruitment, it is merely one way of applying concepts that already exist. All that is needed is for us to consider the possibility that proteins and humans might share certain features of socially cooperative behavior. That the metaphor (PROTEINS ARE HUMANS) is an effective means of allowing us to do this does not show that the metaphor is indispensable, when a simile might have worked just as well.

A quite different argument for the indispensability of metaphor was offered by Richard Rorty, who suggested that the effect of metaphor is not especially linguistic at all, that novel metaphors are like any other surprising experience—say the observation of a platypus—that cause us to revise our current "theories so as to fit them around the new material" (Rorty 1991, 167). This account of metaphor's contribution to scientific discovery is like the popular misconception that it was a bump on the head from a falling apple that led Newton to have his brilliant insight into the theory of universal gravitation.[10] Metaphors, under this scenario, are little more than a cause of insight, and so could only serve a heuristic function. However, the effectiveness of metaphor at getting us to notice certain similarities between disparate things is not so haphazard. Were it so, devising a good metaphor would be a matter of chance experimentation, whereas the history of scientific metaphor suggests otherwise. The creator or proponent of a novel metaphor usually has some analogy in mind, although the task of finding further unintended analogies may be left for others.

Ultimately, I don't think it really matters much whether we say metaphors are absolutely indispensable from scientific theory or not, for I am interested in how science actually gets done, not in how some theoretically ideal version of science might be done in some logically possible world. I want to understand the scientific theory of cell and organismal biology we do in fact have, not a logically purified or rationally reconstructed version. That being said, I will take up this more purely philosophical issue in the next chapter when I consider the implications of all this metaphor use for the questions of scientific realism and objectivity.

In favor of the perspective account of metaphor, we can say the following: it seems a rather natural way of accounting for why metaphor is especially interesting and important within scientific thought. It allows us to see (i.e., consider) something we are perhaps familiar with in a new way, revealing or suggesting new and interesting features that it has or may have; and in this way, it suggests novel avenues of research and experiment. Or a good metaphor may allow us to see something we are unfamiliar with in a more familiar way, likewise suggesting further research to follow up. But a really good metaphor will also be borne out by helping to make successful predictions and to discover objectively real similarities between the source and target domains. In this important respect, the use of metaphors in science to facilitate analogical reasoning and model building is a cognitive activity assessable by rational and objective criteria. If the metaphor fails to help scientists detect relevant analogies between the two systems, it is likely to be dispensed with. As Mary Hesse noted, for particular scientific questions not just any metaphor will do (Hesse 1966, 161). A good metaphor must get at some deep structural isomorphism or similarity between the two systems.[11]

However, as Colin Turbayne noted: "A good metaphor is also a beguiling thing. Once it is understood and accepted, one sees the thing illustrated through new spectacles that, when worn for a while, are hard to discard" (Turbayne 1970, 102–3). Indeed, we have seen that a common criticism of the cell theory was that it led researchers to see cells with clearly defined boundaries where none existed, for instance in syncytia (continuous masses of protoplasm with multiple nuclei). Likewise, we should ask, if we think of metaphor's role in science as providing us with a kind of seeing, what sorts of things about metaphor might we not see if we choose to think about metaphor in this way? What features will be pushed by the perspectival metaphor to the periphery of our mental vision? For to look at something from one perspective means not seeing it from alternative perspectives.

The perspective account emphasizes the connection between knowing and seeing. The "knowing is seeing" trope is a very old and compelling one, and the original cell concept was, in an important sense, also intensely visual (think of the reliance on microscopes), because what the concept of the cell did, as used by Schleiden and Schwann, at least in part, was to provide a search image. It encouraged other observers literally to *look* for morphological features associated with clearly defined cell walls. In the development of cell theory, metaphor is the third lens through which we have investigated and understood living things. But as John Dewey and other pragmatists have tried to remind us, the "knowing is seeing" trope promotes certain assumptions or prejudices about knowledge.[12] Dewey spent a lot of time criticizing what he called the "spectator theory of knowledge," the idea that knowing is a kind of passive mental seeing (Dewey [1920] 1954; [1929] 1960). Why, Dewey asked, should we restrict our epistemological model to just one of our five senses? Pragmatists like Dewey urge us to pay attention to our other modes of attaining knowledge about the world, for instance through active experimentation and manipulation of objects and phenomena around us. Similarly, Ian Hacking (1983) has insisted that philosophers of science should not forget that science has two distinct but compatible functions: representing and intervening (i.e., experimenting). Following up on this suggestion, we might propose that we try taking our tactile sense (or sense of touch) as a model for our understanding of how metaphor works in science, and say something like this: that a good metaphor allows us to *grasp* a subject in a better way. And, in fact, we do often talk about "grasping a concept" (in German the verb "halten" means both "to hold" and "to regard"). But what advantage is there to this form of talk? Well, perhaps if we spoke of metaphor as giving us a better grasp of things, we might be reminded that knowledge is not all abstract and conceptual, that some is also tangible and practical. There is knowing *that* and knowing *how*. Science involves both representing how things *are* and intervening in how things *go*. This brings us to the other major account of what metaphors contribute to science.

4. The cognitive tools meta-metaphor

Metaphors are sometimes also described as conceptual tools, cognitive devices, or instruments of thought or speculation.[13] Tools, of course, come in a variety of forms and functions. Some enhance our perceptual abilities, allowing us to see things we cannot normally see (e.g., microscopes, telescopes); some are instruments of analysis (e.g., surgical scalpels or a high-speed centrifuge) with which we dissect and take things to pieces; some are tools of measurement, allowing us to compare one thing against another (e.g., rulers, balances); and some allow us to get a better grasp on things (e.g., pliers, wrenches, forceps, tweezers).

But in addition to the physical interaction with the material world, scientific investigation also involves the intellectual dissection and rearrangement of ideas, concepts, and other representations of the "external world." Various metaphors serve as different metaphorical versions of all the sorts of tools and instruments previously mentioned for the purpose of performing intellectual or cognitive work. Analysis is the conceptual operation whereby something (a physical object, a process, or a conceptual representation of either) is taken apart and its more basic component parts identified and studied. Thinking of the biochemical activity of cells as organized according to the metaphor of an electronic circuit, or as a chemical laboratory or factory whose materials and operations are organized in space and time according to various stages of assembly, helps scientists to discern the relevant components of the system as a whole, and provides them with hypotheses to guide experiments and investigations. In this sense, the metaphors serve as tools of analysis, like scalpels with which to dissect the complex wholeness of living cells. In their discussion of how a metaphor can give rise to a theoretical model, Soskice and Harré say "the model gives rise to, 'spins off' a matrix of terminology, which can then be used by the theorist as a probative tool" (Soskice & Harré, 1995, 304). The attribution to cells of internal signaling pathways and a circuit-like logic regulating their biochemical activity would seem to have served as such a probative tool, which has encouraged scientists to dissect the cell (both conceptually and experimentally) into the relevant components corresponding to the purported signals, receptors, effectors, switches, and so on.

This particular example also demonstrates what I would argue is a fourth role of scientific metaphors: as tools or instruments of intervention and manipulation, which result not just in changes in our perception of the thing to which they are applied (cells and organisms, or how we understand them), but also in alterations to the very material being or nature of the reality in question. In this case cells, through use of the chemical factory and signaling circuitry metaphors, have been and are in the process of being reengineered, rewired, and reprogrammed, to be more efficient factories or to correct various diseases and pathologies. In a sense, then—an admittedly metaphorical sense—these metaphors have functioned as a kind of experimental tool or technology.

These metaphors facilitate modes of experimental analysis and intervention—they are not primarily a view of life—for they are, objectively speaking, almost certainly wrong and misleading—but they do provide a grasp or handle on the subject by suggesting how to proceed in a reductionist and analytic way so as to reengineer cells to fit our own images and ends of what they could be (cf. Landecker 2007). In this sense, the perspective and lens metaphors are too passive, while the tool metaphor for metaphor emphasizes the pragmatic, interventionist, and experimental aspect of the scientific enterprise.

Insofar as the metaphors are intended to function as these kinds of interventionist tools (like pliers, wrenches, and soldering irons), the question whether they are accurate or true "pictures" of reality is not really to the point. The chief concern is whether they are adequate to the task. Just as one doesn't ask whether a set of vice-grips is true when all that "matters" (note the pragmatic-physicalist metaphor here) is whether they allow one to get a good grip on the thing in question. Of course, one can object that unlike a set of pliers, a model or metaphor can be expected to "work" only on the condition that it faithfully represents or corresponds to at least some of the relevant features of the object to which it is being applied. But I wonder how enlightening such talk of our models representing and corresponding to the way things really are actually is, once we note the reliance of such language on visual metaphors and the spectator theory of knowledge. Consider, for instance, that we cannot actually, i.e., literally, see signaling pathways or networks. Biologists construct circuit-like diagrams and gene-regulatory networks in an attempt to "map out" the causal relations among the various molecules and parts of the internal cell milieu—not *every* causal relation but only those regular ones that seem to matter for the phenomenon in question. They are considered adequate if they result in reasonably accurate predictions and allow biologists to intervene successfully in the behavior of the cell and organism. The comparison between scientific models and maps has become a popular one (e.g., Giere et al. 2005; Giere 2006; Winther forthcoming), and just as a useful map need not be an entirely accurate visual representation of the area of which it is a map, so can models be of significant utility while failing to be very accurate pictures.[14] Models may be taken as representations of real-world systems used for "surrogative reasoning" (Contessa 2011), and they can be said to be "faithful" representations if there is relevant structural similarity between the vehicle and the target; but for the pragmatic purposes of explaining and predicting, a model need not be made more faithful to the target in order to be explanatorily or predictively successful. As a matter of fact, adding more complexity or verisimilitude to the model may hinder its utility. And so with the metaphors that often lie at the center of a theoretical model, relevant structural similarity between the source and target domains is what matters, not superficial dissimilarities. Scientific models and metaphors are not meant to be final descriptive truths of the world but pragmatically useful instruments.

For these reasons, it may be more helpful to think of metaphors as tools than perspectives or lenses (although lenses are a type of tool). However, as Evelyn Fox Keller has commented with regard to the difference in approaches between engineering—which seeks to build and change the world—and pure science, which is typically presumed to be more concerned with understanding: "certainly it is impossible to disconnect the workings of hand and mind" (Keller 2009b, 294). My suggestion is that neither approach is ultimately more correct than the other, but that they are complementary ways of understanding the diverse activities and objectives of scientific practice.

5. Alternative meta-metaphors

What other metaphors have been used to describe the function or nature of metaphor? Max Black suggested that a metaphor works as a kind of "filter" through which the object of the metaphor is passed, such that a set of properties commonly associated with the subject ("a system of associated commonplaces") of the source domain are applied to the subject of the target domain (Black 1962a, 39–40). In saying that man is a wolf, the metaphor encourages us to note human traits that are commonly associated with wolves (41). The result, Black explains, is that "the wolf-metaphor suppresses some details [of humans], emphasizes others—in short, *organizes* our view of man" (41). This is also evidently a version of the visual or perspective metaphor, as Black's next few lines makes clear.

"Suppose I look at the night sky through a piece of heavily smoked glass on which certain lines have been left clear. Then I shall see only the stars that can be made to lie on the lines previously prepared on the screen, and the stars I do see will be seen as organized by the screen's structure. We can think of a metaphor as such a screen and the system of 'associated commonplaces' of the focal word as the network of lines upon the screen. We can say that the principal subject is 'seen through' the metaphorical expression—or, if we prefer, that the principal subject is 'projected upon' the field of the subsidiary subject" (41).

The idea that metaphors organize our understanding of a thing is also suggested by Eva Feder Kittay, who describes the action of metaphor as "rearranging the furniture of our minds" (Kittay 1987, 314ff.). This metaphor is more active and interventionist than the passive viewing or perspective metaphors, but it suggests that metaphor results in little novelty beyond rearranging the ideas already present. According to the interaction thesis made popular by Richards and Black, metaphor is an important source of new ideas; it creates new furniture and accent pieces in the mind.

In a new afterword included in the 2003 reissue of *Metaphors We Live By* (Lakoff and Johnson 2003), there is a section titled "Metaphors for Metaphor" (252–54), where the authors discuss the evolution of their own thinking about metaphor in terms of different metaphorical models. Their first model, they explain, was based on what they call the Mathematical Mapping metaphor. This metaphor led them to understand metaphor as a kind of mapping from one domain into another (hence the terms "target domain" and "source domain"). Its drawback was that it did not account for the creative aspect of metaphor, whereby one thing becomes quite a different thing in virtue of the metaphor. A better metaphor to capture this, they felt, was the Projection metaphor, according to which a metaphor involves a "projection" of the features of one thing onto another, like projecting one slide onto a second, thereby giving new features to the second. But this metaphor implied that all the features of the source domain would be projected onto the target domain, and this certainly is not the case with metaphor (when we say "man is a wolf," we don't suppose that humans have tails and four legs). Metaphors work by making only partial mappings or projections from source to target domain.

Lakoff and Johnson state that by 1997 they had adopted what they call the Neural Theory of Metaphor (which is not itself a metaphor but a neurological theory of how metaphor is processed in the brain). The neural theory recognizes that many metaphors are "complex," meaning that they arise from more simple primary metaphors that are directly grounded in our embodied motor-sensory experiences in the world. So, for example, the conceptual metaphor MORE IS UP is based on our experience that an increase in volume is often correlated with an increase in height or altitude. From this we get metaphors such as "things are looking up," "prices are rising." Lakoff and Johnson think the extension of such primary metaphors into more complex metaphors can be explained through the Neural Theory of Metaphor developed by the computer scientist and artificial intelligence researcher Srinivas Narayanan. The conflation that occurs in the use of a metaphor between two separate conceptual domains is paralleled, according to this theory, by a physical mapping or connection between two or more neuronal clusters in the brain. "Neurons that fire together wire together," Lakoff and Johnson write (2003, 256), and so they proclaim that "Metaphor is a neural phenomenon" (256). The conceptual maps and projections of their earlier theories of metaphor become physical neural maps between different regions of the brain and nervous system.

But although this may provide an account of how our embodied experiences in the world ground our primary conceptual metaphors, it must be noted that this Neural Theory of Metaphor is quite different from any of the other accounts discussed. It is their attempt to provide a scientific account of metaphor that isn't itself reliant on metaphor. But it really doesn't do the same sort of job or answer the same sorts of questions as the metaphorical accounts of metaphor. It really is concerned with quite different questions. It doesn't explain *what* metaphor does for our thinking, but *how* metaphors are processed in our brains.[15]

That being the case, I will leave it aside and turn back to our alternative accounts of metaphor, all of which are themselves based on metaphors about metaphor. Davidson (1984, 245) has said that "Metaphor is the dreamwork of language," by which he meant the interpretation of a metaphor is an open-ended process with no static, fixed meaning; and in a similar vein, Nelson Goodman (1979) spoke of metaphor as language's version of "moonlighting," i.e., a word or phrase working an extra job.

Some metaphors, it has been noted, play an especially important or influential role in organizing how we think about a subject in quite general or broad strokes. The machine and agent metaphors discussed in this study are such. These sort of metaphors which have a very general influence on the way we end up understanding a subject, like cells, have been called "root metaphors" (Pepper 1942) or generative metaphors (Schoen 1993). These tend to be metaphysical assumptions or "world views" rather than the basis for specific testable scientific hypotheses, and for that reason Black (1962b, 240–41) preferred to call them "conceptual archetypes," to distinguish them from either metaphors or models. I prefer Blumenberg's term "background" metaphor (*Hintergrundmetaphorik* 2010), for I think of the machine and agent metaphors as frequently lying in the background of our thinking, almost unobtrusively and subtly exerting their influence, much in the way that a background stage-setting in a theater creates in us expectations about the kind of story we are about to see even before the actors enter and begin speaking their lines.

What other ways of conceiving metaphors there may be is probably only achievable by experiment, by trying out other metaphors. Why not a transposon metaphor? Metaphors are transposons or transposable elements (like those in the genome) that shift units of genetic information from one locus to a novel locus, at once maintaining an established "meaning" while also creating new "meaning" in the process. Or consider crosstalk between semantic pathways of discourse as a metaphor for the effect of metaphor. There seems no end of possible metaphors to be used to think about metaphor itself. To quote Black (1962a, 39): "I have no quarrel with the use of metaphors (if they are good ones) in talking about metaphor. But it may be as well to use several, lest we are misled by the adventitious charms of our favourites [sic]."

But is there no hope for a literal, objective, i.e., nonmetaphorical description of how metaphors function? Let's see.

6. What about a literal account of why metaphors are valuable in science?

One way of offering a literal account of metaphor is to insist on what Black called the substitution view, whereby a metaphor is just a substitute for an equivalent literal expression (employed merely for entertaining stylistic purposes) or perhaps fills a momentary gap in literal vocabulary (i.e., an instance of catachresis). On such an account, metaphor is equivalent to simile. I am not inclined to dismiss this, for it does seem possible, at least in principle, that we could rephrase a metaphor in the form of a simile. However, I suspect the claim that metaphors are ultimately replaceable with literal language is plausible only while we are thinking of science as a finished product and not of science as a process in the making. To say that scientific metaphors can always be replaced with a more literal equivalent is somewhat like saying that a piece written for piano can be played without the pianist having to play scales during its performance, which is undoubtedly true; but it neglects that not only are the scales an important part of the pianist's training, but more significantly, that the scales influence the very format of the music that is written, as consideration of the difference between European music, based on the major and minor scales, and non-Western music like Chinese and Indian, which is based on the pentatonic scale attests.

So I would submit that it's not that we couldn't give a literal account of how metaphors function in science, it's just that such an account would be rather limited and short on insight. It's possible, just not very interesting.[16] Because metaphor is by nature such an abstract thing or process, we have difficulty thinking about it except by comparison with or analogy to other more familiar experiences and phenomena; hence the nearly irresistible urge to speak about how metaphor functions in metaphorical terms.

Note that even the accounts offered by cognitive psychologists and linguists like Gentner and Lakoff end up resorting to metaphor to explain how metaphors do what they do ("transfer of semantic fields" and "neurons that fire together wire together"). Attempts to replace metaphor with some other literal account (with a different sort of tool) will fail, I suspect, to perform the same kind of work.

Is a scientific metaphor just an alternative expression for a simile? It is certainly true that interesting scientific metaphors of the sort that we have focused on in this study (those that are candidates for what Boyd called "theory constitutive metaphors") help us draw analogies, and analogical reasoning basically involves saying that one thing is like another thing. However, metaphors are more open-ended than a mere simile—they are not exhausted by any particular analogies that actually do get drawn. Another important difference is that while a simile suggests only that X and Y are *like* one another *in some way*, metaphor suggests a deeper identity: that X *is* Y or X *is an instance of the sort* Y. Everything is like everything else in *some way or other*, so it's no big deal to say, "A cell is *like* a factory." But to speak of the cell metaphorically as a factory is to suggest some deeper underlying similarity that supports identification at some level between the two or the inclusion of the one concept (cell) as a specific instance of another, more general, category (factory). This is why metaphors are such powerful tools of analogical reasoning: they provide the springboard for creative "mental leaps" (Holyoak and Thagard 1995). But it is also why they are potentially more dangerous and misleading than similes.

The pioneer of modern metaphor studies Ivor Richards put it well—using, of course, a metaphor—when he insisted, against any literalist account of metaphor, that

> words are the meeting points at which regions of experience which can never combine in sensation or intuition, come together. They are the occasion and the means of that growth which is the mind's endless endeavour to order itself. It is no mere signalling system (Richards 1936, 131).

Richards's point is that when we use metaphor we don't simply hold signs for two different things up next to one another in our heads for mere comparison; something more transformative and almost magical happens, by which one thing is transmuted into a very different sort of thing.

7. What about "dead" metaphors?

Well, talk of magic may be excused in the case of fresh or novel metaphors, but what about long-familiar metaphors from which the magic, shall we say, has faded? When a metaphor has become so familiar that we no longer recognize it as a metaphor, it is said to be dead.[17] In the case of a dead metaphor, a shift in the original meaning of a term has occurred, so that "there is no need to consult the original meaning in order to understand a dead metaphor" (Pawelec 2006, 118). The notion of the cell has become a dead metaphor in the sense that when we think about the object thereby denoted, we no longer think of the associated properties; we don't think of little prisons, or our bodies as having a honeycomb structure. We saw in the first chapter that because of these original associations, there were frequent attempts in the nineteenth and early twentieth centuries to replace cell with various technical and nonmetaphoric terms such as plastid, living matter, bioplasm, energid, yet none were able to catch on. However, once people's associations of what they understood the term "cell" to mean had changed sufficiently from the original idea of an empty space surrounded by a solid wall, its replacement was unnecessary. The metaphorical connotations associated with the term cell had died, leaving only the relevant properties sanctioned by the science of the day. The term cell now just means "the fundamental unit of life," and no one thinks of an empty space sealed off behind a rigid and inflexible wall.

So then, are those things we see under the microscope *literally* cells? Well, yes and no. Consider the legs of a chair or table. Do chairs literally have legs or do they only possess legs metaphorically? Is it not literally true that a typical chair has four legs? I think most of us would say yes, but if pressed we might admit that they aren't *real* legs, they're *chair* legs. And likewise, I think we might be inclined to say that, while it may be true that living things are composed of cells, it's *biological* cells they're made of, not *real* cells, obviously. What this illustrates is that through extended use and linguistic development, a concept originally of one genus (legs or cells) may differentiate or speciate into distinct conceptual species (*chair* legs, *biological* cells).

Polysemy refers to the possession by a word or phrase of two or more different though related meanings. The word "bed," for instance, can mean the place where people sleep or the bottom of a river. The meaning of "riverbed," however, has its origin in the former meaning, as it refers to the place where the river "lies" (another example of polysemy and metaphor!). Polysemy can result, therefore, when a word or phrase is used in a novel metaphor. The linguists Bowdle and Gentner (2005) propose a "career of metaphor" hypothesis, according to which a novel metaphor begins as a cross-domain comparison between two separate concepts, but as the metaphor becomes conventional, it is processed as a categorization. On first meeting the term cell factory, for instance, people are likely forced to seek relevant similarities between the two domains in order to make sense of the expression. But as it becomes more familiar, as it does for those scientists working in the relevant fields and reading the specialist journals, a cell just becomes another instance of the more general category *factory*. As we saw in chapter 2, some cells have literally become real factories because they have been reengineered to function as factories producing various bioproducts— they are, at least in part, human artifacts. That, however, is a special case of a metaphor becoming a literally true description. Except for those cells that have been intentionally manipulated to overproduce certain proteins and other products, I think we can safely say that talk of cells being chemical factories is still metaphoric, although increasingly a conventional metaphor with which many people, especially students, are familiar. What should we say, though, about talk of cell signals? Is this still metaphorical or is it now literally true that cells *communicate* with one another by means of *signals*, that these signals are frequently *transduced* and *processed* along *pathways*, *networks*, and *genetic circuits*? How are we to decide whether an originally metaphoric expression has become dead or literal? And is being dead the same as being literal?

To speak of a metaphor being dead may be misleading, for it suggests that the metaphor is no longer vital or vibrant and so no longer doing any work. But in fact, some so-called dead metaphors are more powerful or vital than when they were novel metaphors that still wore their figurativeness on their sleeves—for they work surreptitiously, without our notice; they have become "invisible" (Keller 2002, 209). Take, for instance, the term "mechanism," which originally evoked connotations of machinery and mechanical springs and cogs, etc., but now is used to mean any causal arrangement between a phenomenon or explanandum and its explanans (e.g., the mechanism of inheritance, natural selection as the mechanism of evolution) (see Ruse 2005). If we become too accustomed to looking for mechanisms to explain how cells function, should we not be on guard that we don't assume that any complex set of interactions among proteins within a cell, for instance, are part of some mechanism, and that they must therefore be serving some important adaptive function? In other words, might not the habit of talking about mechanisms, and more particularly of programs and circuits and subroutines, lead us to assume adaptations where there are only contingent or accidental processes? Consider the metaphors of signal pathway or signaling/genetic circuit. Should these metaphors become so familiar that those employing them cease to recognize them as metaphors, they could become a hindrance to further understanding. This is the reason that some people counsel against the use of machine and information metaphors in biology (e.g., Pigliucci and Boudry 2011; Boudry and Pigliucci 2013; Nicholson 2013). Some apparently dead metaphors are not ineffective or inactive. Perhaps it would be more fitting to call them "zombie" metaphors, for like the zombie they can still get around and walk among us; they are not lacking all vital signs— they are, to speak more accurately, "undead."

Metaphors are often introduced into a scientific discourse encapsulated in shudder or scare quotes. This marks them as potentially slippery or dangerous beasts that need to be contained and handled with caution. With time and extended use, these concepts first introduced as metaphors may manage to shed, outgrow, or escape their scare quotes and become free to move around, exercising their powers at liberty and autonomously of our critical attention. It is precisely when they are so liberated—when they have become conventional or dead metaphors— that they sometimes enjoy their greatest vitality. Some metaphors, like that of signal or information in biology, become entirely commonplace without it being clear whether they qualify as dead or not.[18] Perhaps we could call them "transformed" metaphors, in comparison to transformed cancer cells that have taken on a new phenotype and behavior; and like transformed cells, these familiar metaphors have a way of metastasizing and spreading (a word used metaphorically is like a cell undergoing an epithelial-to-mesenchymal transition?). Black (1979) calls metaphors that are supposedly no longer active or vibrant "extinct." But perhaps better is to call them "vestigial," since they still exist in our speech and thought but no longer have the (or any) metaphorical function they originally served. Cell, for instance, is still a metaphor but a nonfunctional one, as we no longer recognize it as a metaphor, nor does it suggest to us any of the original associations of honeycomb, monk's rooms, or other small enclosed spaces.

So what about dead metaphors? Are they ever literally true? The career of metaphor hypothesis suggests they mark a category extension, which, if supported by relevant structural or relational similarities, we might want to say are true or literal. Some conventional metaphors of cell and molecular biology, e.g., cell signaling, cell factory, protein machines, kinase switches, are far from dead. Whereas a truly dead metaphor, like cell itself, may be literally apt because it has shed all of its original associations borrowed from its source domain, these and other terms discussed here continue to evoke their original metaphorical associations, they just have become so familiar that they no longer sound strange or problematic to those who have become accustomed to using them.

It was perhaps for these reasons that the French endocrinologist Claude Kordon proclaimed that "no metaphor is really explanatory," that they only reflect "the cultural references through which we have been conditioned to decipher reality" (Kordon 1993, 96). Metaphors, he seems to suggest, deal in appearances, not reality. But it is worth noting that not all scientists believe scientific explanations must be parsed in entirely literal language. Richard Lewontin, for instance, has said that scientific explanations, "if they are not to be merely formal propositions, framed in an invented technical language, but are to appeal to the understanding of the world that we have gained through ordinary experience, must necessarily involve the use of metaphorical language" (Lewontin 2000, 3). It is time, then, to consider this important question.

8. Are metaphors ever explanatory?

Opinions about this question will of course depend on one's understanding of what it means to explain something, and not surprisingly philosophers have not reached an agreement about this. According to the covering law model proposed by Hempel (1965), a scientific explanation is a deductively valid argument wherein the explanandum or conclusion is derived from a combination of a statement of initial conditions and a universal law of nature (collectively called the explanans). According to the inductive-statistical theory (Hempel 1965), an explanation is an argument in which the explanans (premises) make the explanandum (conclusion) likely. Both these accounts, which have been subject to serious criticism, are incompatible with metaphor unless we permit statement of the "laws of nature" and/or initial conditions to be parsed in metaphorical terms; in which case, since metaphors are, strictly speaking, false, Hempel would likely say we fail to have a true explanation (Hempel 1965, 338). (An irony here is that the very notion of a law of nature is based on a metaphor. See Zilsel 1942; Lehoux 2006.)

The unification account advocated by Friedman (1974) and Kitcher (1989) holds that to explain a phenomenon is to show how it can be derived from another established theory or explanatory framework, thereby unifying our understanding of the world and how it works. This is in principle more metaphor-friendly, since the extension of an analogical situation between two or more originally disparate phenomena frequently occurs through the use of a metaphor. All these accounts of explanation tend to draw on examples from the physical sciences, and so emphasize the explanatory significance of laws (often expressed in the form of mathematical equations) or very general argument patterns. They tend to downplay the significance of causes, largely through a Humean-empiricist distrust of the notion of causal powers as providing an objective ground for the regularities we observe in nature. Others, such as Harré (1972) and Salmon (1984), however, argued that to explain a phenomenon is to describe the causal process that brought it about, and more recently the centrality of causal mechanisms for scientific explanation has formed the core of the "new mechanist" accounts discussed by people whose views have been informed by the biological and life sciences. William Bechtel (2006), for example, has drawn attention to the fact that explanations in cell biology are almost always formulated in terms of the concept of a mechanism, not deductions from universal laws. And like the concept of a law, the general concept of a mechanism is rooted in a metaphor, as frequently are more specific examples of mechanisms.

A mechanism basically consists of a number of parts standing in some causal relation to one another such that they jointly bring about some particular phenomenon (Glennan 1996; Machamer, Darden, and Craver 2000). As defined by Bechtel (2006, 26):

> A mechanism is a structure performing a function in virtue of its components [sic] parts, component operations, and their organization. The orchestrated function of the mechanism is responsible for one or more phenomena.

So what does a mechanistic explanation look like? I will provide brief sketches of two recent examples, one fitting quite closely our everyday understanding of a machine-like mechanism, the other a little less so. The first arises in attempts to explain cell motility, how eukaryote cells manage to crawl about on substrates like other cells, the extracellular matrix (as they normally do during developmental cell migration), or a glass microscope slide. Many eukaryote cells exhibit what is called amoeboid motion, similar to the slow oozing motion of amoebae by the extension of pseudopodia, with the exception that the leading edge of many animal cells advances by the protrusion of a ruffled sheet of cytoplasm called a lamellipodium. The eukaryote cell contains a dynamic cytoskeleton composed of actin, microtubule, and intermediate filament polymer molecules that functions as a flexible and shifting internal framework providing the cell with structure and shape. The cytoskeleton is also implicated in the cell's creeping motion along a substrate. When filamentous actin (F-actin) is polymerized at the leading edge of a cell, it creates a force against the cell membrane, resulting in the protrusion of a lamellipodium; at the same time, the opposing force exerted on F-actin by the cell membrane creates a retrograde F-actin flow. If the cell membrane was not attached to the substrate, this would only create a change in shape of the cell, with no actual forward motion. The cell membrane makes contact with the substrate through focal adhesion (FA) points based around transmembrane mechano-receptors known as integrins. It is believed that by establishing a physical connection within the cell between the dynamic polymerizing F-actin and the FAs, the cell transfers the frictional force created by attachment of the FA to the substrate so as to move itself forward. As FAs at the rear of the cell are disassembled, the cell is able to crawl forward in what is described as an inchworm fashion. But in order for this proposed mechanism to work, there must be some way to engage and disengage the F-actin to the FAs. The existence of such a mechanism to couple the intracellular F-actin flow to substrate-attached focal adhesions is called the "molecular clutch" hypothesis (Lin and Forscher 1995). (See fig.5.1A.) Until

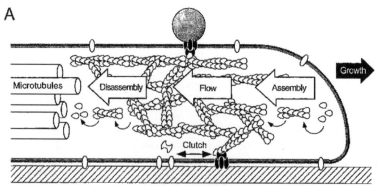

FIGURE 5.1 Molecular clutch hypothesis for cell motility. (a) Early model. (From Chi-Hung Lin and Paul Forscher, "Growth Cone Advance Is Inversely Proportional to Retrograde F-Actin Flow." *Neuron* 1995, 14:763–71, fig. 8, with permission of the author and Elsevier.) (b) Recent model, showing the "clutch" mechanism engaged and disengaged. (From Margaret L. Gardel, Ian C. Schneider, Yvonne Aratyn-Schaus, and Clare M. Waterman, "Mechanical Integration of Actin and Adhesion Dynamics in Cell Migration." *Annual Review of Cell and Developmental Biology* 2010, 26:315–33, fig. 3, with permission of the author and Annual Reviews, Inc., permission conveyed through Copyright Clearance Center, Inc.)

B

a

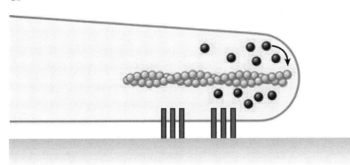

b Disengaged

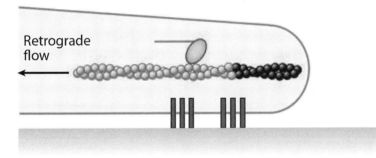

Retrograde
flow

c Engaged

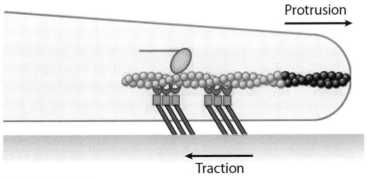

Protrusion

Traction

FIGURE 5.1 (*continued*)

recently, the nature of this "clutch mechanism" was not well understood. But a recent study by Clare Waterman at the NIH and her collaborators has identified the protein vinculin as a molecular clutch (Thievessen *et al.* 2013). As the authors explain: "Despite extensive evidence for the molecular clutch hypothesis . . . it is unclear which molecules engage F-actin retrograde flow to integrins in FA. Thus, it is not known how F-actin engagement regulates F-actin organization and FA maturation and dynamics" (164). They report (on the basis of gene knock-out and biomechanical experiments) that vinculin is a force-bearing F-actin binding protein that forms an essential component of the clutch mechanism by regulating the formation and maturation of FAs (see fig. 5.1B). Cells in which the vinculin gene has been deactivated can still move, but in an unorganized and less efficient manner. The authors report that "together with our demonstration that vinculin promotes traction force at FA, this suggests that vinculin mediates conversion of forces generated in the cytoskeleton that drive retrograde flow into traction force on the ECM [extracellular matrix] during FA maturation" (173). Here then is an example illustrating how the analogy and the metaphor of a mechanical human artifact suggested to researchers an analogy providing a possible explanation for the function of a natural biological system, and how the metaphor has helped to increase the investigators' understanding.

My second example deals with a slightly less machine-like sort of mechanism. It involves signal transduction in neurons. Here is how the authors express the explanation in question in the article summary:

> Activity-dependent CREB phosphorylation and gene *expression* are critical for long-term neuronal *plasticity*. Local *signaling* at Ca_v1 *channels triggers* these events, but how *information* is *relayed* onward to the nucleus is unclear. Here, we report a *mechanism* that mediates long-distance *communication* within cells: a *shuttle* that *transports* Ca^{2+}/calmodulin from the surface membrane to the nucleus. We show that the *shuttle* protein is γCaMKII, its phosphorylation at Thr287 by βCaMKII *protects* the Ca^{2+}/CaM *signal*, and CaN *triggers* its nuclear translocation. Both βCaMKII and CaN *act* in close proximity to Ca_v1 *channels, supporting* their *dominance*, whereas γCaMKII *operates* as a *carrier*, not as a kinase. Upon arrival within the nucleus, Ca^{2+}/CaM *activates* CaMKK and its substrate CaMKIV, the CREB kinase (Ma et al. 2014).[19]

I have added the italics to indicate those terms that are metaphors. There are seventeen distinct metaphor types here (with twenty-one metaphor tokens): eleven verbs and six nouns. Neuronal plasticity is important for learning and memory, and CaMKII is a protein kinase[20] considered to have an important function in learning and memory pathways. In order for neurons to be responsive to new experiences, signals conveying relevant information must be transmitted from the outer membrane of the nerve cell to its nucleus, where gene expression responds in the appropriate way for memory and learning to occur. Once carried to the cell nucleus, the Ca^{2+} signal activates the transcription factor CREB (cAMP response element-binding protein) to bind to DNA and switching on or off specific genes. Ca^{2+} binds to Calmodulin (CaM) within the cell to form the Ca^{2+}/CaM complex, but CaM lacks a nuclear localization sequence (an amino acid sequence that acts as a "tag"), which is required to shuttle it to the nucleus, and consequently it requires a "carrier" to get it there. Ma *et al.* provide evidence that the γ form of the Calmodulin kinase (γCaMKII) is that "shuttle."[21] So, have the authors offered here an explanation of how the Ca^{2+} signal gets from the nerve cell membrane to the nucleus, where gene expression results as a response to events outside of the cell? They certainly seem to think so, as they write, "This mechanism resolves long-standing puzzles about CaM/CaMK-dependent signaling to the nucleus" (Ma *et al.* 2014, 281). I suspect most scientists would also agree that they have offered *a potential* explanation, even if it does not ultimately turn out to be *the* correct or complete explanation. Now what would this explanation look like if it were expressed in strictly literal language? We could in theory replace all the metaphoric nouns (*plasticity, channel, information, mechanism, shuttle, carrier*) with wholly new terms lacking any prior semantic associations and substitute a more prosaic term such as "causes" for all the more descriptive action metaphors (*signals, triggers, relays, communicates, transports, protects, acts, supports, operates, activates*), but would it retain its cognitive content? Would it be improved by such a literalization of its language? Improved for what or whose purposes?

One could, of course, dispute that the terms I have singled out as metaphorical are in fact metaphors at all—perhaps they began as metaphors, but now that the entities and activities they denote are well understood scientifically, aren't these instances of dead metaphors? Haven't the meanings of terms such as signal, information, transport, shuttles expanded over time so that these uses are now quite literal? I will return to this specific question about dead metaphors and explanation in section 9 below, but first I want to make some concluding remarks about scientific explanation.

Attempts to provide a purely formal account of explanation in science (in terms of logical or structural relations between sentences or statements) have fallen short, as have attempts to identify some feature shared in common by all instances of what we would be inclined to recognize as legitimate instances of scientific explanation. This has led to attitudes of pluralism, pragmatism, and contextualism about explanation(s). I believe that if we pay attention to how scientists actually go about explaining things to themselves and to their peers (as in the examples above), we might be inclined to expand on a suggestion made by Ron Giere (2006, 60) about how to understand the nature of scientific theorizing more generally. Whereas traditional accounts have construed science as a two-place relation between the world and theories about it, Giere recommends thinking of it as a four-place relation between a scientist or group of scientists S, some aspect of the world W, a representation (or model) of that aspect R, which is used by scientists for some specific purpose P. Likewise I propose that rather than think of explanation as a two-place relation between a real-world phenomenon (P) and an account of it (A), such that "A explains P" in some absolute sense, we should think of explanation as a four-place relation involving a scientist (S), some aspect of the world (W), a representation or model or metaphor (M) of W, which is used by S for some specific purpose (P). Scientific explanation then will be construed as follows: S uses M to represent W for purpose P, where P might stand for "to explain why W happens" or "to understand W." How closely M must fit W, how similar the model or metaphor must be in order to be a good or successful explanation, will depend on the scientist(s) and their intended purpose. What may count as good enough for one scientist or group of scientists, or for one purpose as opposed to another, may vary from situation to situation (cf. Keller 2002, 300–2). This does not rule out the possibility that the group of people who find the explanation adequate for their purposes might ultimately include the entire scientific community. If this is an adequate account of scientific explanation, then we can say, "Yes, metaphors can be explanatory or can be the basis of an explanatory model or account." Yet we may suppose that those metaphors that help to identify some significant structural, relational, or causal similarity between the two terms are most likely to be favored by scientific researchers.

9. Polysemy, explanation, and understanding

Attempts to formulate a truly universal account of scientific explanation capable of capturing every instance of what we would intuitively count as an explanation have failed. As mentioned above, this has led some to adopt an attitude of pluralism about the concept of explanation. Van Fraassen (1980), for instance, argued that explanation is an essentially pragmatic activity rather than a formal relation between different types of sentences expressed in observational and theoretical language. An explanation, he suggests, is an answer to a why question ("Why did this occur?" or more generally "Why is it the case that P?"), and assessment of whether something counts as an explanation, or how good an explanation it is, will vary from one specific context to the next.[22] Peter Godfrey-Smith (2003) agrees that there is no universal form that all explanations must assume, and that what counts as an explanation varies from one scientific discipline to another, a position Godfrey-Smith refers to as *contextualism*. Having listened to cell and molecular biologists argue about how "deep" one must go in order to explain a phenomenon (need the explanation include details about individual molecules of amino acids or the gene sequences of DNA implicated in their synthesis? or is a higher level of description sufficient and acceptable?) this contextualist account sounds right to me.[23]

If one insists there must be some feature—if not necessarily a formal or logical one—that is shared by all instances of explanation in virtue of which we call them all explanations, one might suggest they all advance our *understanding* of a phenomenon. But because it seems such a subjective matter, understanding has not commanded a great deal of respect from philosophers of science, many of whom are happy to leave it for the psychologists to deal with.[24] What is it that provides that peculiar "aha!" moment when we feel we now understand something? And is it always a reliable indicator that we now have proper or thorough knowledge of the thing in question?

It is frequently claimed that the cognitive significance of metaphors is that they allow us to understand something unfamiliar by describing it in terms more familiar. So on this account, to explain something is to make it seem familiar. But as Hempel (1965, 257) noted, familiarity is neither necessary nor sufficient for a good explanation. Many of the things for which we might seek explanations as to how they work are already quite familiar to us (I am familiar with my computer and that it can send and receive e-mail through the wireless access to the

Internet in my home, without having any understanding how any of it works, despite attempts by more tech-savvy acquaintances to explain IP protocols, etc., to me); and the principles appealed to in science to explain things are often unfamiliar and sometimes even outright strange (appeals to gravitational attraction at a distance, space-time curvature, or exchange of graviton particles to explain why things fall to the earth, for instance). Something more than familiarity is required, it seems, but what? We have seen that in the context of molecular and cell biology, explanations are typically articulated through description of causal mechanisms rather than by appeal to high-level theoretical laws or generalizations, as is more common in physics. Analogical reasoning is important in the sorts of context within which cell and molecular biologists operate, as it allows them to extend structural relations from one domain where they are known to apply to another where they might also apply. These structural relations may involve mechanistic analogies, or they may be cast as analogies with social relations among agents in a larger system, as illustrated by the examples of cell and molecular sociology.

It is the facilitation of analogical reasoning by metaphor, which allows for the discernment of similarity between dissimilar phenomena, that leads many thinkers to associate metaphors with scientific models, analog models, to be specific. But not all similarities are of equal importance. Superficial similarities, for instance of mere property ascriptions, are not very helpful for understanding why things happen as they do. Many—e.g., Black (1962b), Hesse (1966), Gentner and Jeziorski (1993)—have insisted on making a distinction between mere metaphor and a proper scientific model, the difference being that a model must be based on a real and reliable analogical structure or isomorphism existing between it and the thing modeled. While a metaphor may serve as the original inspiration for a fruitful scientific model, it must, as Kuhn (2012) said of paradigms (with which they have also been closely associated), be "articulated." This involves conceptual clarification of the metaphor so as to make specific predictions and its application to novel problems or phenomena.

So to say that cancer is the result of one "renegade cell" (Weinberg, 1998), for instance, is to offer more in the way of a description than an explanation for neoplasia and tumor growth. What it does do is to highlight the essentially social nature of normally well-behaved metazoan cells within the strict order and rule-governed system of the "society of cells" (the body, the organism), and thereby frames the phenomenon to be explained in a particular way. Devising an explanation why a cell might begin to behave in a disorderly fashion that is detrimental to the well-being of the organism of which it is a part requires further probing into the internal workings of the cell (while keeping in mind that causal influences originating from both its immediate tissue microenvironment and its broader ecological environment outside of the organism may also be relevant). This has led scientists to seek understanding of the causal mechanisms governing normal cell physiology (in particular the growth and division cycle) in the hope that they will be able to discover some altered or malfunctioning component(s) in that regulatory causal system. Weinberg and other proponents of the somatic mutation theory have worked for the last couple of decades within the metaphorical space created by the language of computer circuitry and wiring diagrams to look for explanatorily suggestive analogies of how alterations to cell signaling pathways caused by genetic lesions lead to reprogramming of various integrated regulatory circuits, and ultimately in cancer (Hanahan and Weinberg 2011).

More recently, Weinberg (2014) has expressed a more restrained judgment of the success to date of this model and its assumption that beneath the diversity of cancer phenomenology there lay a relatively simple and linear logic-like program complicated only by the occurrence of positive and negative feedback loops. Despite optimism of systems biologists to explain and indeed predict how "the complex machine—a human cell—works" (Weinberg 2014, 270), Weinberg reports that the explosion of data in the -omics era confounds simple intuition of its meaning, that the "coupling between observational data and biological insight is frayed if not broken," and that scientists currently lack the "conceptual paradigms" and computational strategies needed to deal with the complexity (271).

Weinberg's opinions notwithstanding, it would seem that many proponents of systems biology do regard the use of terms such as *signal*, *information*, *program*, and *circuit* as legitimate extensions of their original electrical and computer engineering meanings. One could, therefore, make an argument for the legitimacy of metaphor in scientific explanation on the basis of polysemy—that is, when the meaning of a term has been extended in such a way that it now has multiple related meanings, each of which is considered by a community of language speakers to be literal. A conventional or dead metaphor, in other words. So when scientists describe the response of a cell to an external signal that binds to a membrane-bound receptor, which then transduces it via second messengers along a signaling pathway or network, are they at this point in the history of the development of scientific knowledge still speaking in metaphors or in terms that have simply come to mean precisely and literally what the scientists say they do? Some may be inclined to answer in the affirmative, while others may ask whether the fact that a community of language speakers considers a term polysemous or now literal is on its own sufficient to show that this judgment is justified. As Rothbart (1984, 608) notes, "Any metaphor is potentially literal depending on the general compliance of the community of a speaker." Is there not a difference, however, between the simple fact that some originally metaphorical term, e.g., information or molecular switch, becomes polysemous for some community of language speakers (and of what proportion of the whole body of those considered experts?) and there being good evidential reasons for a community treating such a term as polysemous? A legitimate judgment that a once-metaphorical term is now a conventional or dead metaphor or literal description ought to rely on the successful revelation of a pertinent isomorphism between the two domains (a true homology as opposed to a mere analogy?).[25] Although some biologists talk of biological information and signals as being computed by cells as if these terms were as legitimate within biology as they are in computer science (e.g., Brenner 2012), others worry that such talk is potentially more misleading than informative (Pauwels 2013). This is an expression of a realist worry: are these *really* signals, are cells *really* computing? Because computer engineering and electronics technology is now such a rich and complex domain, it has great potential to be both a far-reaching and informative guide as well as a very misleading one. To compare a cell to a computer is to open a very large box of possibilities, not all of which can be expected to strike deep and important similarities (Hesse's "positive analogies") between the two domains.

As legitimate as such concerns are, it is not immediately clear that this should disqualify all explanations that include metaphorical language from being truly or properly explanatory. Consider again the example of the molecular clutch (Thievessen et al. 2013). All would agree, I think, that the molecular complex composed of vinculin and talin is not a *real* mechanical clutch. But what the scientists employing it do maintain is that it works enough like a clutch mechanism to be heuristically insightful and to help them understand how a cell manages to coordinate the motion of its internal cytoskeleton and its external focal adhesion points so as to move in a crawling fashion. So of course, they might concede, talin is not a *mechanical* clutch, it is a *molecular* clutch. And if one insists on asking, "But is this molecular clutch talk *truly* explanatory?," we might legitimately wonder whether something of the "No true Scotsman" move is being made (Flew 1971, 47). This is an informal fallacy whereby one who has made a universal assertion (e.g., "No Scotsman reads that newspaper") responds to an offered counter-example ("MacDuff reads it") by redefining the terms in an ad hoc fashion just to exclude it ("No *true* Scotsman reads that paper!"). It is a form of special pleading.

On the other hand, it is surely true that not every instance of metaphor in science constitutes a successful explanation of some phenomenon. Take, for instance, an example used by the cancer biologist Sir David Smithers (see chapter 3). Smithers was critical of the thesis that cancerous tumors arise from so-called "cancer cells," a central thesis in the somatic mutation theory. In making the case that, aside from leukemia and other blood disorders, most cancers present as abnormal tissue organization (i.e., tumors), Smithers remarked that "cancer is no more a disease of cells than traffic jam is a disease of cars. A lifetime's study of the internal combustion engine will not explain it" (Smithers 1962, 495). This is a highly effective analogy for illustrating his contention that it makes little sense to look for the cause of a higher-level phenomenon like cancer at a level lower than a tissue, since an individual cell can no more "have cancer" than a single automobile can constitute a traffic jam. Yet as effective as this metaphor is for conveying and clarifying his *understanding* of cancer, Smithers's analogy could not be said to offer an *explanation* of cancer. It could at most explain, if correct, why attempts to understand carcinogenesis by means of investigation of individual cells (and the genetic mutations they undergo) would prove unsuccessful, or at least incomplete. The metaphor or analogy between a tumor and a traffic jam highlights certain organizational similarities between the two (disorganization of the component units), but it does not isolate a similarity of relevant causal import: cars do not create a traffic jam by replicating themselves in an uncontrolled fashion, they just pile up for a variety of reasons. The analogy seems better suited perhaps for thinking about the cause of atherosclerosis, but even then, unless the reasons why blood cells get clogged in the arteries are similar to why cars get stuck in traffic, we might hesitate to say the analogy properly *explains* the natural phenomenon rather than provides some suggestive avenues of experimental investigation to pursue in the hope of finding more specific and appropriate causal details. If, on the other hand, we were to suggest that some traffic jams are caused by the malfunction of traffic signals, we may be able to point to an interesting and analogous causal mechanism: disrupted or abnormal signaling between cells *is* considered an important causal factor in the initiation of neoplasia and tumor growth by the proponents of the Tissue Organization Field Theory of cancer (Sonnenschein and Soto 1999).

If, however, we agree with Kordon that metaphors only reflect the cultural lenses through which we have learned to see reality, we must discount the possibility that the subjects related by a metaphor ever share enough structural similarity or isomorphism for the metaphor to provide any real understanding of the one in terms of the other. Hempel (1965, 440) too was of the opinion that even an analogical model that establishes a real syntactic isomorphism between the statement of laws in two separate domains "explains nothing" (441), although he did allow—as did Duhem (1991, 95ff.)—that they can be heuristically useful for discovering the real laws that do. Insofar as both their accounts of explanation assume the general adequacy of the covering-law model, we may count them as of less than crucial significance within the quite different context of cell and molecular biology, and perhaps of the life sciences more broadly. In consideration of Darwin's theory of evolution by means of natural selection, for example, Michael Ruse (2005) has insisted on the key significance of the culturally and historically contingent metaphors of selection and of organismal adaptations as machine-like contrivances serving specific functions. Darwin would not have arrived at this wonderfully successful account of the organic world, Ruse argues, had he not approached it through these metaphors and analogies. In light of this, Ruse concludes:

> Given the success of the science, it would seem silly if the philosopher—in the name of some kind of epistemic purity—were to object to the way that things are. In the real world, science—the best science—reflects the human beings who create it (Ruse 2005, 300).

But this does raise the question of how we measure success. We might legitimately ask, for instance, whether the computer engineering metaphors that have become so entrenched in cell and molecular biology have been so successful. And successful in what regard?

If we are agreed that a naturalist perspective on science is the proper approach, then a mark of a metaphor's or analogy's success must be tied to what its users can do with it. Does it help them to make progress in their attempts to better understand, to predict, and/or to manipulate the system in question? And what other criterion could we use, given that this sort of success is our only means of evaluating whether any representation is true or empirically adequate? Some worry that naturalism with respect to science leaves us without any normative ground to stand on. But in addition to looking at what progress scientists are able to make with a metaphor (what phenomena they believe they understand better because of it, and what fruitful suggestions for new phenomena and questions it has delivered), we can also ask whether it has any significant negative effects. And this does provide some normative bite to the naturalist approach.

On this count, we find lots of agreement among both philosophers and scientists, that one must be careful not to mistake the model for the thing being modeled. For a metaphor or analogy may become so entrenched that it leads us to "construct . . . data that conform to it" (Stepan 1986, 133), or, as Chew and Laubichler (2003, 53) put it: "biology's metaphorical abstractions all too easily become concrete objects and substitute for specific, describable processes." Weigmann (2004) summarizes the potential deficiencies or dangers of metaphors as the following: (1) They may lead to neglect of aspects of reality that don't fit the metaphor; (2) overemphasize those aspects that do; and (3) suggest unintended and misleading connotations. It is for these reasons that people (e.g., Pigliucci and Boudry 2011; Ball 2011) like to quote the cyberneticists Arturo Rosenblueth and Norbert Wiener that "the price of metaphor is eternal vigilance."[26]

Massimo Pigliucci and Maarten Boudry (2011) call for the rejection of all "machine-information" and engineering metaphors (Boudry and Pigliucci 2013) in biology as highly misleading and additionally for providing unintended support for the Intelligent Design movement. Not only are machine-information metaphors "deleterious for science education," they argue, but "they actually misdirect or partially derail thinking about what sort of research programs biologists ought to carry out and how" (Pigliucci and Boudry 2011, 13). They echo the concern of the nineteenth-century English pioneer in cell biology Sir John Goodsir that "there is nothing . . . which has more retarded science and philosophy, and the kindred subjects on which human reason has been employed, than the introduction of terms with conventional meanings" (Goodsir, quoted in Cleland 1873, 257). Such claims are familiar now with respect to the discipline of genetics, where talk of "master molecules," "genetic codes," and "programs" has come under repeated criticism. Similarly, Nicholson (2013; 2014a) strongly criticizes what he calls "the machine conception of the organism" and calls for its abandonment. But he also concedes the heuristic utility of machine metaphors for efforts to understand the function of parts of organisms and of cells; the error, he believes, is in taking the machine metaphors as adequate for understanding the cell or organism as a whole.[27]

But Nicholson, Pigliucci, and Boudry seem to assume that science (or more restrictedly, biology) strives for one uniquely adequate description or account of things, rather than trying to achieve multiple intellectual and technological objectives. Science, however, is not one, nor even is biology—there is rather a diverse array of biological and biomedical sciences with diverse questions and projects, and so no one perspective may be adequate for them all.

As for the concern that metaphors lead research astray, one can reasonably respond that metaphors serve to generate hypotheses, and through the process of conjecture and refutation, biologists can become aware of inadequacies in the models/metaphors to improve the reliability and adequacy of their explanatory accounts. Even an entrenched metaphor that comes to form the core of a paradigm may, as Kuhn suggested, serve to highlight anomalies and thereby provide an important impetus to conceptual progress. The demands of creativity and scientific progress suggest that scientists will continue to experiment with new metaphors and analogies (cf. Paton 1992; Keller 2002; Moskowitz and Aird 2007; Loettgers 2013). There is no other way to know in advance how far a metaphor is adequate but to push it to its limits. And in this respect metaphors are no different than any scientific instrument: the user must be aware of any imperfections in its design and results, of any artifactual errors it may introduce. Compare early microscope lenses with their defects of spherical and chromatic aberration, which gave rise to epistemological or "reflexive" concerns about whether and how they could be reliably employed to generate knowledge (Schickore 2007). Any instrument—material or conceptual—must be used with proper appreciation of its limitations and systematic errors, as well as its advantages. Metaphors and the models they help to devise may need to be calibrated and refined to become more reliable, trustworthy instruments. But if we would reject an instrument because of its imperfections, its failure to provide us with perfect fidelity to objective truth—think of our own preferred natural instruments, our eyes—science would be not just seriously hobbled but outright impossible.

10. Metaphors as visions of research programs

One last significant function fulfilled by metaphors needs to be mentioned: that is their status as a vision or aspirational goal of what a research program or project hopes to achieve. As Boyd (1993, 489) notes: "Precisely because *theory-constitutive metaphors are invitations to future research*, and because that research is aimed at uncovering the theoretically important similarities between the primary and the secondary subjects of the metaphors, the explication of these similarities and analogies is the routine business of scientific researchers" (emphasis added).[28] A metaphor may provide a scientist or team of researchers with a motivating goal or objective for pursuing a particular program of research; it may serve to collectively organize their individual efforts, which is no insignificant thing. I have in mind the metaphors of biomachines, BioBricks, cell factories, rewired or reprogrammed cell circuits. It is probably no coincidence that the visionary use of metaphor tends also to be amenable to the form of visual metaphor. This aspect of metaphor in science and technoscience is closely similar to Nelkin's (1994) *promotional metaphors*, those figures of speech having a largely rhetorical and persuasive function of winning over scientific collaborators and financial investors.

In this regard, the perspective metaphor is doubly metaphorical: for it's not just that a metaphor allows us to *see* a subject from a new perspective, it also provides us with a *vision* of the subject, not in complete isolation, but in relation to other objects or subjects. We see how the thing relates to other things, and we get thereby a roadmap of how to proceed in further investigation. Metaphors can then play a unifying role (Holton 1995, 280) by weaving together different aspects of our experiences into a coherent *Weltbild*. Such grand world-view metaphors are equivalent to Black's (1962b) archetypes, Blumenberg's (2010) background metaphors, and Pepper's (1942) root metaphors. This is an important feature of metaphor's metaphysical or system-building capacity. Metaphors promote research programs, specific methodological approaches and techniques—which speaks again to their prescriptive component. By highlighting certain questions, they inevitably obscure or downplay others, and in doing so favor certain techniques and methods of investigation over others.

Conclusion

Metaphor plays a significant cognitive role in science, beyond its obvious rhetorical and pedagogical functions. Metaphor is of obvious great significance in the context of discovery. It has immense heuristic value for how scientists conceive of the world and the problems they find worthy of investigation (and those they ignore or fail to notice), and it influences (both positively and negatively) how they carry out their investigations, as well as how they gather and interpret their data. But metaphor can also play an important cognitive role in the development of explanations of natural events and phenomena. It does this by facilitating analogical reasoning, through the development of models that highlight structural, relational, or causal similarities between the phenomenon to be explained and other aspects of our experience of which we have better understanding. Although a novel metaphor says something that is, strictly speaking, false, through its extended use the meaning of a term can be expanded so that what was once metaphorical becomes literal. Other terms originally used metaphorically may become so disassociated from their original implications that their conventional meaning or connotation no longer relies on the properties of the source domain; they have become dead metaphors. However, many metaphors, I have argued, exist in some limbo state between being fully live and completely dead. Such conventional metaphors retain still some of their original metaphorical connotations even if many of those who use them fail to recognize this. Rather than referring to these as dead metaphors, I suggest calling them zombie metaphors, because we tend to use them unconsciously, as if they—or rather we, under their influence—are in some kind of trance-like state.

I have also recommended that we be more mindful of the metaphors we use when engaging in second-order discussion about why metaphors are of value in science. The most common of these, what I have called the perspective metaphor, states that metaphors provide a new or useful perspective of a subject that allows us to see it in an illuminating way. This is a passive understanding of what metaphors do, the unreflective use of which can prevent us from seeing metaphor from other perspectives: for instance, as tools of various kinds that allow us not just to passively view a subject, but to probe and to dissect, and in some cases to refashion or rebuild so as to change not only the way we see the thing, but the very nature of the thing itself.

Commentators on the role of metaphors in science have tended to suggest that a given field of research is dominated by one all-important master metaphor. But I have tried to emphasize how researchers often use multiple, even contrary, metaphors in the run of their investigations. In this regard, too, metaphors are like tools and instruments, serving specific needs in specific contexts and at specific stages of scientific inquiry. At the risk of pushing a metaphor too far, we might think of scientists' use of different metaphors as comparable to the different magnifying lenses of a microscope, allowing them to zoom in at different levels of the object of their study as their interests and the demands of inquiry require. The scientist may use a social or agential metaphor one moment to describe the behavior or function of a cell or protein, and then a machine or mechanistic metaphor to describe or explain how it manages to perform that function. Only if we believe there is one uniquely correct perspective or level of analysis from which scientists are to speak about the world should this be unsettling to us. In the next chapter, I consider the repercussions of science's deep reliance on metaphors for our general image of science and the knowledge it provides of the world around and within us.

6 The Instrumental Success of Scientific Metaphor:
Putting the scientific realism issue into perspective

Science aims to give us, in its theories, a literally true story of what the world is like.

Bas van Fraassen (1980, 8)

Science aims at fruitful metaphor and at ever more detailed structure.

Ernan McMullin (1984, 35)

1. Realism, literalism, and objectivism

Metaphors can be thought of as tools or instruments of
scientific inquiry that assist us in making sense of the
world and of ourselves. They are conceptual tools—
which may mean that they are tools only in a metaphorical sense, depending on whether or not we want to consider this a dead metaphor—but their importance for
the scientific enterprise is made real enough through the
great success with which scientists employ them. The use
of metaphor by scientists is *instrumentally successful* in a
way similar to how microscopes or stained slides of organic material are successful in revealing to us things we
hadn't previously seen (i.e., recognized) or understood.
But if we ask whether various types of instruments or
techniques, say fluorescence microscopy or cell fractionation, show us things as they "really" are, the answer is
no. They distort and alter the natural state of living cells;

they produce artifacts. And yet these artifacts do make accessible to us important properties of the objects being studied. In the same way metaphors may not provide us with an objective account of the world *as it really is*, independent of how we talk and think about it, but they can nonetheless help to reveal to us important features of the objects and processes to which they are applied and to understand them in relation to other areas of science and common experience.

One of the big questions in philosophy of science has been whether the claims made by science (especially about what lies beyond our immediately observable experiences) are true. Does science describe things as they really are? This is the problem of scientific realism. It is well known that a hypothesis or theory can be instrumentally successful in the sense that it makes more or less accurate predictions about the world (or our experiences and observations of it), without being true. The Ptolemaic-geocentric hypothesis of the universe is a classic example of an account that made plenty of successful predictions about the positions of celestial objects (construed relatively to our position here on Earth), yet turned out to be false on many important details. Likewise, in medicine there have been occasions when physicians have successfully treated patients using a treatment that they did not properly understand. So both philosophers and scientists have been understandably reluctant to take the instrumental success of a hypothesis as a wholly reliable sign that it gets at the truth of things.

According to an influential statement by Bas van Fraassen, scientific realism is the thesis that "*Science aims to give us, in its theories, a literally true story of what the world is like; and acceptance of a scientific theory involves the belief that it is true*" (1980, 8).[1] Now if scientific realism means aiming at a literally true story of what the world is like, then science's reliance on metaphor is a problem for the realist; not only because metaphors appear to be literally false, but because, as we have seen in cell biology at least, it is common practice for scientists to use multiple metaphorical descriptions of the world that seem to be inconsistent with one another. The relevance of metaphor for the realism issue goes beyond the traditional question about whether we can know unobservables (which is the typical concern of philosophers of science), because metaphor is also standardly used in science's accounts of observable things. Not all cells are microscopic after all: consider the egg of a chicken or a nerve cell from the neck of a giraffe, which can run up to 4.5 meters (15 feet) in length; and if one believes that things seen with the aid of a microscope count as observable, then we must ask whether cells really signal to one another, commit suicide, and so on.[2] According to van Fraassen's expression of realism (which is by no means peculiar within the philosophy of science literature), the occurrence of so much metaphor in cell biology must mean one of two things: (1) either this science is not yet complete (the metaphors need to be replaced by more objective or literally true language) or (2) the anti-realist is correct that science fails in its attempt to give us a literally true account of the world.

Another option—known as *instrumentalism*—insists that scientific theories are not really intended to offer true accounts about the ulti-mate reality of the world at all. They are, rather, instruments used by us humans to make predictions about and to control events in the world as things appear to us. In that case, truth or falsity is no more relevantly ascribed to a theory than it is to a screwdriver or other tool. What matters is whether it works for us, to help us achieve any number of different human goals. Instrumentalism seems a fitting interpretation of some scientific theory and activity, e.g., the bioengineering attempt to rewire cell signaling pathways or to create factory-like BioBrick components; and metaphors, as previously noted, are powerful tools of intervention that drive many of the objectives of biotechnology and synthetic biology. But metaphor is not restricted to the intervention side of the scientific enterprise. As Ian Hacking (1983) emphasized, science is concerned with both *representing* nature (which entails providing ac-curate descriptive accounts) and *intervening* in it (which includes both experimentation and the kind of technoscience represented by synthetic biologists). In short: science seeks both to *represent* and to *reconstruct* nature (e.g., cells).

So what are we to make of metaphor's role in science's attempt to de-scribe the way the world is? Does it count against scientific realism? At one end of the spectrum we have the thesis that the world has an inher-ent structure and is packaged into units of its own devising, which are completely independent of our attempts to describe it. This, as Stathis Psillos (1999, xix) states, is a metaphysical thesis or stance essential to the position of scientific realism.[3] At the opposite pole is the thesis that the world has no inherent structure whatsoever except what we humans impose upon it ("Man is the measure of all things"). This radical social constructionism is an equally metaphysical thesis. It is no longer as pop-ular as it seemed to be even a decade or so ago, when scores of books and academic papers were proclaiming everything from genes to laws of nature to be socially constructed rather than objectively discovered.[4] There seems to be a consensus now that the conclusion to be drawn from all the heated debate about the Science Wars is that the truth lies someplace in the middle: Science is both an activity of creative invention and of discovery (a conclusion I think everyone recognized already, though to varying degrees). But to appreciate properly the insights of each side of the debate for the question of scientific realism, we need to distinguish carefully between two separate positions that I believe are commonly conflated. These I will call literalism and objectivism.

2. Distinguishing literalism from objectivism

To highlight the difference between the two, consider the statement: "This dandelion is a weed." If used to describe an unwanted flower on my lawn, it would express a literal truth, if in fact I did not desire it to be there. That's a literal use of the term "weed." (If I said of those ugly commercial signs with the moveable bright letters that they are "weeds in the urban landscape," I would be saying something metaphorical.) But from the fact that I can make literally true statements using the term "weed," it does not follow that it can be used to express *objective* truths about the world, because, aside from the intentions of human property owners, "weed" is not what is called a natural kind term. "Weed" is a useful socially constructed category for us to use, but plants do not come objectively preordered by nature itself into weed and nonweed varieties.

Consider now van Fraassen's statement of scientific realism and ask what it would mean to say that the cell theory was literally true? First, I think we would have to say that for a good portion of its history, it was not literally true that all living things are composed of one or more cells, because the notion of a cell was, until sometime around the turn of the twentieth century perhaps, still a fresh metaphor; and as a metaphor says something that is literally false, it could not be literally true. But at some point in history, cell became a dead metaphor and so a literal term used to denote the fundamental unit of life. From that point on, we could say that the cell theory was literally true (while keeping in mind that some biologists did and still do have reservations about its full adequacy). But whether or not cell is a natural kind term picking out part of the universe's own structure, rather than just a convenient social construction used by humans for organizing our experience of living things, that is a different question.[5]

It is in fact almost a truism or tautology of modern biology today to say that all living things are made of cells.[6] Nature, on the other hand, as the evolutionary biologist and unicellular specialist Thomas Cavalier-Smith said on one occasion, is messy and non-Platonic.[7] Given any particular living thing, it will not always be clear whether it is one cell or several. Still, we can affirm that it is made of cells of some number and thereby accept the cell theory as a convenient "system of artificial memory" with which to tie up a bundle of facts, to use T. H. Huxley's phrase (1853, 249). This is a similar point to the one made by Nancy Cartwright about the laws of physics, that they are not quite true of particular real-world events or systems but are rather defining principles of certain types of models that we humans use to understand segments and aspects of the real world (Cartwright 1983). Likewise, metaphors play an important role in making the world make sense to us—recall Woodger's comment about the satisfaction found from identifying a repeating motif in a wallpaper pattern.

But more importantly, I want to make the case that "literally true" does not mean the same thing as "objectively true." For a word to be used in a literal sense just means for it to be used in its "normal," non-metaphorical, sense. There need not be, nor is there, any guarantee that a term used literally describes the world "objectively," as it "really is," independently of us humans and our peculiar ways of experiencing and talking about it. In fact, the only clues we ever have that our accounts or descriptions of things are "objective" is that they prove to be instrumentally successful, which is to say, technically powerful and empirically adequate, across a wide range of experiences and circumstances that are accessible not just to one human subject alone, but to anyone who is able to position themselves in the relevant perspectives.

The idea that "there is exactly one true and complete description of 'the way the world is'" is what Hilary Putnam called "metaphysical" or "external" realism (Putnam 1981). Metaphysical realism assumes that we should understand truth as a correspondence between our conceptual scheme (our language, concepts, beliefs, etc.) and a wholly independent external world. Ron Giere prefers to call this "objectivist realism," but rejects it all the same in favor of what he calls "perspectival realism" (Giere 2006), a thesis that shares with Putnam's "internal" realism the conviction that we can only understand true statements as being made from within or internal to some language/conceptual scheme/perspective or other. Giere's perspectival realism is informed by close attention to the actual practices of science and its significant reliance on the construction of models that scientists use to construct partial, though accurate and useful, representations of selected aspects of the world. Giere (2006, 5) maintains that the most we should be willing to say of any bit of good science is that, "according to this highly confirmed theory (or reliable instrument), the world seems to be roughly such and such." We should resist making the unjustifiable objectivist claim that "this theory (or instrument) provides us with a complete and *literally* correct picture of the world itself" (6, italics added). I think, though, that Giere is confusing here the question whether a scientific theory or statement is *literally* true with the question whether a scientific theory or statement is *objectively* true. Only the second should give us pause, for we can and do make literally true statements in science and elsewhere all the time, e.g., "You are now reading a book," "The Sun is a G-type main-sequence star," or "Fibroblasts are animal connective tissue cells of mesenchymal origin." Assenting to the *literal* truth of these sentences need not render us guilty of metaphysical hubris of the kind Putnam and Giere rightly caution us against.

Theodore Brown, in his book-length study of metaphor in science, also seems to miss this distinction when he takes issue with "the strong realist position . . . that we humans can aspire to attain a literal, universally true understanding of nature" (Brown 2003, 187). Brown adopts Lakoff and Johnson's Conceptual Metaphor thesis that many of our most important abstract concepts are metaphorically based on our experience as embodied beings in the world, and from this Brown argues for a form of realism he calls "embodied realism."[8] In expressing this position, though, it becomes clearer exactly what he intends to be rejecting: "Embodied realism denies that there is a single, absolutely correct description of the world" (187). But again, one can reject objectivist or metaphysical realism (an "absolutely correct description of the world") while allowing that we can aspire to literal truth about the world.

Although science does manage to say literally true things about the world, we should recognize its claims are only ever true from the perspective of our beliefs, language, and models, etc. So I believe the present study of science's reliance on metaphor supports Giere's perspectival realism. But, to be clear, that conclusion is not that there is no objective reality "out there," or that science is just a collection of socially constructed narratives. What is really being denied is that there is any good reason to believe that science can be expected to arrive at *one uniquely* correct objective account of reality that is parsed *in nature's own terms*.

Nor am I arguing that all language is metaphorical, right down to its roots—for the simple reason that the concept of metaphor requires its contrast of literality in order to have any cognitive content (or as we often say metaphorically, "to make sense"). Again, we can and do make literally true statements, e.g., "Human cells are of the eukaryote variety" is literally true. "Embryo cells must *decide* what kind of specialized cell-type to become" is almost certainly metaphorical. And some statements can even manage to be at once both literally true and metaphorical, as we saw with the John Donne-inspired statement "No cell is an island." But to say that we can make literally true statements about the world is not the same as saying that those *literally* true statements are *objectively* true in the sense that they are made in nature's own uniquely correct language and vocabulary. For, to repeat, whether a term is used *literally* or not is a matter of the conventions of a group of language speakers. It is a matter of pragmatics (in the linguistic sense), not metaphysics. Whether a term is *objectively* true or correct in the external realist sense is, as Putnam said, a matter of metaphysics, and not one that can be easily resolved, so far as I can see. But none of this means that we cannot retain a form of realism that is accessible to humans, what Putnam (1992) once called "realism with a human face," or what Giere (2006) more recently calls perspectival realism.

There are those, however, who wish to defend external realism and the thesis that there exists a real world that is wholly independent of us and how we think, talk, or feel about it. John Searle, for instance, writes, "If there exists a real world, then there is a way that the world really is. There is an objective way that things are in the world" (Searle 1999, 15). But this raises two questions. First, what does Searle mean by "*a way*" and "*an* objective *way*" the world is? Does he mean an account, a description? What would be the source of this "way" that exists independently of beings who would provide the account? And second, does it really follow that "If there exists a real world, then there is a way that the world really is?" Consider the analogous bit of reasoning: "If there exists a real past, then there is a way that the past really is/was. There is an objective way that things are/were in the past." I would argue that one can be a realist about the existence of a world independent of us without believing that there is one uniquely correct account of it, just as one can be a realist about the past without believing there to be one uniquely correct account of historical events. Historians, like scientists, may or may not reach an agreement about how best to describe some past event like the French Revolution, for instance, but if they do, it may be just as much a contingent fact about historians as it is of the past itself. They may reasonably expect to reach agreement about basic facts, that an event occurred at this place on this date, yet we may reasonably expect disagreement about how to describe it (was it a proletarian or a bourgeois revolution?), and about its causes and its significance for later events; for even though we may reasonably suppose that the past really happened, what reason have we to believe that it has its own inherent and uniquely correct structure? Granted, natural scientists are not concerned with such obviously social phenomena, but they face similar challenges in their attempt to accurately describe the facts and put them into some kind of coherent narrative and causal framework.

Where does the assumption come from that belief in the independent existence of the world requires belief that it also has its own inherent and unique structure?[9] Searle maintains that the belief that "there is a way that things are in the world independently of our representations" requires no justification, nor that it even makes sense to attempt to provide one, because "any attempt at justification presupposes what it attempts to justify," and that "any attempt to find out about the real world presupposes that there is *a way* that things are" (31, italics added). But isn't this also a *non sequitur*? For if it were correct, wouldn't historians (who presumably do believe the past really happened) have to throw up their hands and admit that, because they fully expect to come to different conclusions about some things, given the rather inchoate nature of the past, their attempt to find out about it was an impossible waste of time? I would like to say that scientists, like historians, may reasonably suppose their efforts are not in vain so long as the object of their investigation is not entirely fickle; but that nature has its own inherent singular "way," to which their own accounts must correspond in order to count as successful science, this strikes me as an unnecessary presumption.

Well, these are very deep waters, and I'm not that comfortable myself when I can't touch bottom. And some readers will no doubt be asking themselves, what about objectivity? Isn't to give up on objectivist or external realism and the belief that there is a way the world is independent of us to give up on objectivity altogether? To answer that concern, I will attempt to provide a clearer sense of what the notion of objectivity is supposed to mean for a group of language-speakers. And to do that, I suggest it is helpful to look at several key metaphors with which the notion of objectivity has traditionally been bound up.

3. The metaphorical underpinnings of the concept of objectivity

Realism of the external or objectivist kind is itself intimately based on a particular set of metaphors that are intended to motivate the notion of objectivity as a stark contrast to the notion of subjectivity. The motivation I understand and agree with, it's the metaphors used to articulate it that I find lacking. The English term *subjective* comes from the Latin *sub*, meaning "under," and *jacere*, "to throw." *Objective* comes from *ob*, "in the way of," and again *jacere*, to throw (*Chambers's Etymological Dictionary* 1966). Hence, something *objective* is "before us" and is a possible source of resistance or obstruction that can oppose our will;

whereas something *subjective* is, I suppose, "beneath us" and not an obstacle.[10] To be subjective is to see or understand things from one particular perspective, from the perspective or viewpoint of just one subject among many. Being subjective is not inconsistent with having knowledge, understanding, or true beliefs about the world, but it can only ever be partial and incomplete. To correct this deficiency, we attempt to understand a thing from "all sides." This is a principally *epistemic* conception of the subjective-objective distinction. But there is also an ontological conception of objectivity that highlights what are supposed to be the intrinsic features of a thing as they exist independently of a knowing subject. In theory, the two should ultimately coincide as a subject's understanding and knowledge of a thing converges to the thing's real properties and nature.

Let's consider a paradigmatic example. When we find some unusual object—either planted firmly in the ground before us or small enough to pick up in our hand, we are not satisfied to look at it from one perspective or angle alone. We walk around it or turn it over in our hand so as to see it from all angles. We may ask others also to look at the object to confirm our impressions or to discover aspects we have missed or are incapable of detecting. From this type of basic experience and practice, we derive the metaphor of considering an issue or argument from all sides. We seek to achieve not merely an account or description that is accurate from one or even several subjective perspectives, but one we believe would be accurate from the object's own perspective, that is, an account that it would give of itself in its own terms, were it able to speak. Here we have made a move to the *ontological* conception of the subjective-objective distinction. I suspect that we arrive at this conception from our experience of attempting to understand other people (other subjects) by trying to "see things from their perspective."

Seventeenth-century thinkers like Galileo and Descartes made a distinction between an object's primary and secondary properties: the first being real properties inherent in the object like number, size, and shape; while the latter are subjective properties resulting from our perception of objects, such as color, taste, odor, sound, and feel. An objective account would include all and only these primary properties and would amount to a description in the object's own terms or language. That is the experiential source, I believe, of the objectivity and objective view metaphors. But at no point do we humans actually get completely eliminated from the equation. And at no point, as Giere (2006, ch. 4) has argued, do we cease to see or to understand the object from *some perspective or other*.

The idea of objectivity has been further explicated by means of three other key metaphors: "the Book of Nature"; "the view from nowhere"; and a third, which underpins them both, "the God's-eye view." Let's take a quick look at these separate metaphors.

The Book of Nature The "Book of Nature" metaphor—a long-standing trope in Judeo-Christian theology and popularized by Galileo in the seventeenth century (Kay 2000)— encouraged a literalist conception of science as the attempt to translate Nature's language, or that of its designer/creator, into the scientist's own, whether that be Latin, French, German, English, Spanish, Italian, or geometry, as it was according to Galileo (Blumenberg 1981). God is conceived as a divine draftsman (or to use a more modern image, a software engineer) who sits down before actually creating the world to decide what its proper terms will be (its code and ontology) and the laws of its behavior. This again invokes the ontological version of objectivity. A successful or (objectively) true scientific account, then, would be one that made a completely accurate translation from the one language into the other. But without the theological motivation for thinking of Nature as a text created by a language-speaking agent, a major reason historically for believing we could achieve a uniquely correct scientific account is no longer available. We can add to this, arguments for rejecting the idea that the semantic meaning of linguistic units (either words or whole sentences) are well defined enough to make a uniquely correct translation between two languages even a possibility—see, for example, Quine (1969) on the "indeterminacy of translation."

The view from nowhere If to be subjective is to see something from one perspective or a few only, then to be objective is to see the thing from no particular perspective at all (Nagel 1986). The problem with the idea of the view from nowhere, if taken in its literal sense, is that it would be— were it even possible—wholly uninteresting for us humans, for it would be a view *of* nowhere. Consider again our paradigm case of inspecting an object, say a house, from many perspectives: from the front the house will appear one way, from the back another, as well as from each side. And from inside it will, of course, look totally different again. We learn to recognize that none of these singular perspectives is any more correct than the others, that they are all partial pieces of the real object. We combine them to create a conception of the real object—but not into any one unified singular perspective. We do not, for instance, combine them the way we do the individual pieces of a jigsaw puzzle to create one larger two-dimensional picture from the partial two-dimensional images of the multiple pieces. When we complete a jigsaw puzzle, we do not change our perspective, we just arrange the information available to us within that one perspective in a particular way. Our intellectual conception of an object is unlike this insofar as it results not in one unified or privileged perspective but in the balanced holding of a plurality of distinct perspectives, which may be incompatible in the sense that they reveal quite different properties.

In this sense, the view from nowhere is a misleading metaphor for our notion of objectivity. However, the point of the imagery is surely to attain a view from *nowhere in particular*, not necessarily one uniquely privileged account, or even a "view from everywhere" (which would be incoherent or confusing if taken all at once). So rather than seeking any singular viewpoint, we might best think of science as seeking multiple "views from someplace," from which a syncretic view may arise or not, depending on the extent to which multiple perspectives can be fruitfully aligned or simultaneously held. Such an understanding of objectivity is more sympathetic to a pluralism of perspectives rather than one uniquely correct account.

The God's-eye view Traditionally the view from nowhere was not just some special *human* perspective, but that of God the creator. As an infinite being existing outside space and time, God was supposed to be able to see everything in the universe as it truly and objectively is and all at once. God's perceptual apparatus and categorial framework or conceptual scheme could not distort the inherent structure of reality, for they were supposed to be one and the same. Einstein's special theory of relativity, however, showed the assumptions of absolute simultaneity and a privileged frame of reference to be problematic for mere humans, making any hope of achieving the trick of seeing everything all at once and from the one privileged vantage point seem scientifically misguided. The God's-eye view—like the view from nowhere—combines the ontological and epistemic conceptions of objectivity. Both metaphors assume that the world comes already packaged into its own proper units or categories, that there is, as Searle says, a "way" the world is, independently of us. But in what sense is this privileged "way" accessible to us? And is it even required for us to make sense of the scientific pursuit of knowledge?

One of the obvious functions served by a concept of God is to provide some means of resolving conflicts between humans and maintaining social unity within a group. God is supposed to know each person's

innermost thoughts and to serve as an infallibly informed judge on ultimate matters of justice and morality. Such a belief might inspire greater
honesty and cooperative spirit within a group of people.[11] Even among
secular people, there is the belief that when two parties disagree, there
are always three sides: party A's account, party B's, and the Truth. A
third party is often brought in to help resolve disputes, on the condition that they are "objective," meaning that they have no subjective or
personal stake in the outcome.

So the notion of objectivity serves a crucially important social function as a regulative (epistemic) ideal that we should not wish to dis-
pense with. However, is it the case that we also require the metaphorical
(ontological) mythology that has attended the concept of objectivity's
development in order to maintain its positive effects? I think not. If
objectivity is not to be construed as the vantage point or descriptive
testimony of a privileged Super-Agent, then it must be understood in
terms that are thoroughly human, while at the same time avoiding fa-
voritism toward the interests of any particular subject or group. This is
essentially to cease regarding objectivity as a metaphysical-ontological
concept and to conceive of it as a methodological-epistemic ideal. Philosophers of science such as Helen Longino (1990) have proposed re-
conceptualizing objectivity as intersubjective criticism, or replacing "the
view from nowhere" with "the view from everywhere" (e.g., Barker and
Kitcher 2014, 112–13, 161).[12] While this seems like an improvement,
others worry that it results in a relativist muddle, wherein we have no
basis for making critical evaluation of any individual's or group's perspective. I personally would rather speak metaphorically, not of any
one singular view "from everywhere," but of many views from many
perspectives, while keeping in mind that there may be plenty of overlap
among this pluralism of viewpoints, so that we need not worry about
a relativist incommensurability or breakdown of mutually critical and
constructive dialogue, just as we use our experiences from many different perspectives to assemble our conception of one singular object. And
given that we are all biologically human, we surely share many perspectives in common, enough to allow us to critically evaluate our own and
others' beliefs and reasons for holding them.[13]

4. Getting real about realism

If realism requires commitment to the thesis that science aims for one
uniquely true, objective description of the world, then I would not
consider myself a realist. But if realism means having descriptions that

allow us humans—with our particular neurophysiology, our language-dependent cognitive abilities, and the interests and objectives that spur us to scientific inquiry—to effectively navigate through our experiences of and interactions with the world, then realists we should be, but realists of a pragmatic, instrumentalist stripe. *Realistic* realists, I would say, not ones who pine for an impossible God's-eye view of the world or the ability to read the Book of Nature literally in its own language. Nature may in fact permit us to regard it as many books or tell many "stories" about it (to use van Fraassen's ironic term), and to successfully interact with it using many different tools or perspectives. As a matter of fact, the assertion that science aims at *a* literally true account does not necessarily imply that there can be only one unique account, *the* literally true account. It leaves open the possibility of two or more literally true accounts, though raising the question whether multiple literally true accounts could manage to avoid being inconsistent with one another. This is a possibility if none were presumed to offer a complete account, in which case the distinct literally true accounts could be complementary rather than contradictory. Attention to the actual practice of scientists teaches that science proceeds by simultaneously building multiple images of nature that can be overlaid, or partially overlapped in some cases, one on the other, not by insisting from the start on agreement about one unified, literal vocabulary and description.[14] Because science has several aims— explanation and intervention, in addition to true description—there may be no one best solution or account, and metaphor may have a legitimate function in the attempt to achieve any of these ends.

5. The literal-metaphorical distinction

In his expression of scientific realism, van Fraassen included that science aims to give us a *literally* true account of the world because he wished to distinguish his own version of anti-realism (which he called "constructive empiricism") from others like conventionalism, positivism, and instrumentalism. According to these latter positions, a theory's claims about hypothetical and unobservable entities need to be properly interpreted in some "allegorical" or nonliteral fashion (as a metaphor or simile); such claims are either to be translated into talk about purely observable phenomena or to be understood as heuristic tools for making predictions about observable phenomena. Van Fraassen, however, agrees with the realist that if a theory says "There are electrons" (Van Fraassen 1980, 11), then we are to take that statement literally.[15] Van

Fraassen's concern is with the status of language having to do with un-observable theoretical entities and processes. That is not my concern. Mine is how we should take all the apparently metaphorical talk scientists engage in. When they say that "cells communicate by means of signals, which are transduced along pathways that include second messengers and kinase switches that deliver information to specific addresses in the nucleus," how are we to know whether to take this literally or metaphorically? Is it always clear–even to the scientists themselves–when they are talking metaphorically as opposed to quite literally? The language used by scientists and laypeople alike is actually a hybrid of literal phrases mixed with novel, conventional, and dead metaphors. As I. A. Richards remarked:

> The processes of metaphor in language, the exchanges between the meanings of words which we study in explicit verbal metaphors, are super-imposed upon a perceived world which is itself a product of earlier or unwitting metaphor and we shall not deal with them justly if we forget that this is so (Richards 1936, 108–9).[16]

Even if we were to insist that obvious examples of metaphorical expression can be expunged from the ultimate scientific account of the world, we are left with the problem of how to tease apart those bits that are metaphorical from those that are not. The literal-metaphorical distinction, in other words, may be more porous, more a continuum, than a binary gap. It is as if in constructing science it's not always clear which bits are scaffolding and which are parts of the building proper, because the scaffolding can, over time, become a proper part of the finished product. Perhaps rather than the image of science as the erection of a skyscraper, we might think of it as the organic evolution of cells from endosymbiotic partners, in which metaphors, over time, become subcellular organelles firmly and fully integrated into the new living unit. As with endosymbionts, the removal of metaphors from the corpus of a scientific theory may result in a decrease in fitness. Moreover, to pursue this metaphor a bit further, just as understanding an organism's fitness is not only a matter of discerning how well it "corresponds" to its environment, but of understanding its objectives, its goals (e.g., growth, survival, reproduction, perhaps even "flourishing"?), all of which might properly be counted as *values*; so too, science is driven by specific goals and is distinguished from other sorts of human endeavors by a set of distinctive values, among which are those like honesty, openness to criticism, respect for evidence, and proper experimental design—in short, what we collectively call objectivity of the methodological-epistemic kind.

6. Models, metaphors, and instruments

Science, then, is the values-driven attempt to solve problems and puzzles engaged in by human agents ensconced in a historical-cultural context, speaking a language (or languages) they have partially inherited from their predecessors and are partially in the process of adapting and recreating. This is one reason that metaphor plays such an important role in the actual practice of scientific inquiry. The best interpretation of these facts about science is, I would urge, a pragmatist picture that emphasizes science and scientific theory as a human attempt to understand the world and to improve our plight in it, rather than to achieve a godlike and entirely disinterested theoretical reflection of an inherent structure. But it is also one that shares with the realist picture a conviction that we are to take the successes of science seriously, for we want to know why certain metaphors work so well, and why some work better than others for answering certain questions and for achieving certain ends. So far, I have spoken mostly about science's attempt to provide accurate/true/objective descriptions of the world. But science also purports to *explain* why things happen as they do, as well as to *manipulate* and *control* what happens. As we saw in earlier chapters (especially 4 and 5), metaphors have a significant function in these objectives as well.

Explanation & understanding Van Fraassen's expression of scientific realism stated that science aims to give us a literally true "story" of what the world is like. In a way, this is an ironic concession to the essentially human aspect of science, for stories are undoubtedly a peculiarly human creation. A story or narrative is more than just a catalogue of descriptive statements: it involves essentially a selection of descriptions about the world or some topic, and organizes them in a particular way, foregrounding some, relegating others to the background so as to help us to understand its events, to see them as part of some larger pattern or form, or as moving toward some particular *telos* or conclusion. In doing so, we attempt to impose some order on events—to identify some pattern, laws, or generalities: to see the world as more than a string of unrelated facts, as more than just a tale "full of sound and fury, signifying nothing."

The attempt to identify repeatable, predictable regularity and order in the buzzing confusion of events involves the creation of scientific models. A model in the sense intended here is that appealed to in chapter 5, a simplified and abstract representation of some aspect of the world concocted by some person or group of people for some particular purpose. Like maps, with which they are frequently compared,[17] models successfully represent reality to the extent that they permit us to understand, to predict, to manipulate, and to move about in a world that often operates against our wills and frustrates our expectations. The role of metaphor in the development of scientific models, Richard Boyd has said, reveals the "accommodation of language to the causal structure of the world" (Boyd 1993, 483). Metaphors assist us in getting access to relevant causal structures in the world in the absence of clearly and precisely defined terminology, and may eventually be replaced by more precisely defined literal terminology. Likewise, Ernan McMullin used the fertility of metaphors (what I have called their "instrumental success") as an argument for scientific realism. Metaphors can be used to make a "tentative suggestion," he said, about how a causal structure known to function in one area might be extended to a new one, and thereby "illuminate something that is not well understood in advance" (McMullin 1984, 31). When these metaphorical extensions of successful models work, as they often do, we must, according to McMullin, assume that without necessarily being wholly true, they do get at some structural reality. Because metaphors provide the predictive fertility required for scientific progress, McMullin maintained that "science aims at fruitful metaphor and at ever more detailed structure" (35).

The difficulty with scientific language is one of finding the right terms to fit the reality of how the world appears to us to be—what things the world is made of and their properties; and yet the choice is also influenced by the type of questions we are asking. The problem is not as simple, in other words, as trying to find prefabricated nuts (terms/concepts) of the right dimensions to fit bolts that are also already there (nature's "joints" or "furniture," a "ready-made world"). We have to construct or at least isolate conceptually both the nuts and the bolts, and nature itself does not tell us how to do this. So even when we do manage to achieve a fit that works successfully *for our purposes*, we have no guarantee that the nuts and bolts we have identified are really or objectively nature's own. (Are organisms indivisible wholes or compounded of more fundamental units? And if so, what are those units? Organs, tissues, cells, molecules? How should we define these?) What we can say with confidence is that it works very well for us to describe nature in those terms. If the concepts and descriptive terms we create do work for us, if they assist us in making successful and precise predictions and in manipulating aspects of the world or our experiences within it, then we have legitimate reason to believe that they have got hold of some real aspect of the world that is not just of our making. As Mary Hesse pointed out, in science not just any metaphor will do. Successful metaphors are those that get at some deep structural or relational similarity between two domains, they do not trade on superficial attribute similarities. Still, there is always a certain amount of slippage or looseness of fit between our terms/concepts and the language-independent reality we hope to attain knowledge about by their means, even when we intend to be using these terms literally, in their originally intended or nonfigurative sense. This was the motivation for Kuhn's (1993, 537) remark that "if Boyd is right that nature has 'joints' which natural-kind terms aim to locate, then metaphor reminds us that another language might have located different joints, cut up the world in another way."[18]

That the choice of terminology is a contingent one is made especially apparent when the task at hand is one of articulating scientific explanations for why the world works as it does. As discussed in chapter 5, formulating an explanation is a pragmatic activity. To the extent that explanations are ultimately meant to be intelligible to and illuminating for human agents, we should expect not only a plurality of potentially suitable explanations for any given *explanandum*, but also a significant role for metaphors therein. By whatever means it manages to do so, it seems uncontentious that an explanation should help to make an event or phenomenon more understandable to the person or persons to whom it is proffered. But understanding is a curious affair, and what makes intuitive sense, what increases one person's understanding or the understanding of a community of scientists, may vary across historical epochs and across social and cultural contexts. The explanations we find increase our understanding say as much about us (at a particular historical moment) as they do about the things they purportedly explain. The trend since the sixteenth century for mechanistic explanations, and particularly for those appealing metaphorically to the most recent technology of the day (e.g., from telegraph wires in the nineteenth century to electronic computers today), is testament to this. For this reason, we might invert Boyd's remark and say that the use of metaphor in science displays "the accommodation of scientific language to the *cultural* structure of the *human* world."

Moreover, this study has shown how, in cell biology at least, scientists are not always in the thrall of one dominant metaphor or paradigm; they rather frequently make use of instances from the two chief background metaphors, in one and the same paragraph describing proteins, for instance, as operating like machines but also as social agents that recruit others to work with them cooperatively. This tendency for scientists to shift back and forth between such different frames of speech or perspectives should also give us pause in supposing that the end goal of their pursuits will be one canonical, uniquely true account.

Intervening & reconstructing In addition to describing what the world is like and explaining how it works, science also aims at manipulating and intervening in it, ideally so as to improve the human condition. Here, where the goal is less about representing the world than reconstructing it, literally true description may be a less crucial desideratum (though still desirable to some degree), and the role of metaphor may become more significant. As we have seen, metaphors may provide a vision to guide research (e.g., cell factories, rewiring of cell circuits, BioBrick components), in which case the goal is less one of providing descriptions true to nature than of remaking nature to be true to our visions of what it could be. I've argued that in some cases metaphors function like tools, and tools of diverse kinds. The original cell metaphor provided microscopists with a search image, and in that sense we might say it helped them to see living tissue and whole organisms in a new way. Other metaphors, such as signal pathways, circuits, and switches, help biologists to dissect the cell conceptually into components and to experimentally and physically rearrange them; and in this way they function a bit like scalpels and tweezers, allowing scientists to get a grip on the parts and to alter their organization and behavior. If it is correct to regard these metaphors in this way, then concern about whether they can be part of a literally true description of reality is misguided, as it misses the point of how scientists actually use them. We don't ask whether a scientist's instruments are true, we ask whether they are effective and reliable.

It is for these reasons, having to do with the scientific objectives of providing explanations and intervening in the world, that the assumption that science aims to provide a literally true account of the world seems too restrictive. Asking whether science is capable of telling us the truth about what the world is really like assumes two things: first, that the object of science is chiefly descriptive (rather than, say, to be more proactively experimental and interventionist); and second, that the question of realism has to do primarily with theories and what they *say* about the world (rather than what they allow us to *do* in and with it). These assumptions share a similar view or perspective of science as a rather passive activity—it is in fact that of the spectator theory of knowledge, according to which scientific knowledge is supposed to comprise a reflection or mirroring of nature. As an alternative image, the instrumentalist or pragmatist account associated with people like John Dewey encourages us to think of science as a collection of techniques and tools for dealing with our environment. Theories and models are examples of such tools, which can be thought of as attempting to represent aspects of the world (as maps do) without having to raise the question Do they provide us with objective truth of how the world really is, independently of us? What is important is whether they work for the diverse range of tasks we construct them to perform, whether that be descriptive, explanatory, or interventional/manipulative. For a theory or model to work successfully at any of these tasks requires that it correspond with or connect with relevant aspects of reality, but always as those aspects of reality are implicated or reveal themselves in our experiences in the world.

Accounts of scientific realism typically seem to assume a correspondence theory of truth: a true belief or statement corresponds to the facts or reality. And so long as we don't ask what it means for a belief or statement to "correspond to facts or reality," scientific realism looks commonsensical. But if we do ask this question (as the pragmatist insists we should), then we are quite likely thrown back to saying things like "'corresponds' means to be empirically adequate and instrumentally useful, but not just for now and for a few people, but for everyone and right up to the limit or end of inquiry," which is precisely what pragmatists like Peirce understood truth to mean.[19] I know some philosophers will insist they mean more than this—that truth is more than an epistemic notion, that a true belief really does "mirror" or "map onto" the way things really are, independently of us. But these are metaphors, and unlike many of the examples of scientific metaphor discussed in this book, not very useful ones, in my opinion. By contrast, I am frequently struck by how many working scientists, when asked whether they think they are discovering the objective truth about cells or some other aspect of the world on which they work, respond modestly by saying, "No, I'm just trying to build a model that will help me to understand how this system or phenomenon works."

Even if we retain reservations about whether a theory or model gives us reliable truth about the objective and independent properties of some aspect of the world, we may be satisfied that the things it deals with are real, if the model or theory allows us to make accurate predictions and to successfully manipulate phenomena. This is to invoke the distinction between being a realist about theories (about whether their claims about the world are true or not) and being a realist about entities (about whether the things the theories suggest exist do or not) (cf. Hacking 1983). In either case, we should have no reservations about saying that successful science provides access to or knowledge of reality, so long as we keep in mind, as Kant (1992[1781]) said, that it provides us with knowledge of reality insofar as it is a possible object of experience for us humans. In the words of the cell biologist Laurence Picken (quoted earlier): "those only will be disconcerted who believe that science is Absolute Truth. Of that, science has no knowledge" (1960, xxxvii).

If we insist on thinking about science in terms of a visual metaphor of attempting to picture the world, or of producing images of it, then perhaps rather than seeking to provide one coherent image adequate for all of nature and for all purposes (the one uniquely and objectively true account), we should think of it as providing a kaleidoscope of distinct nonrepeated images, all of which strive in their own way to be useful and accurate reflections of nature or, more accurately, of our interactions with it. This is not to say that unification in science by way of more general theories is a bad thing or is not a legitimate goal (in fact, one of metaphor's chief constructive roles in science is to help achieve unification of explanatory principles by means of facilitating analogical reasoning); but it is to suggest that attaining one grand unified theory (even within any particular field of science, let alone science as a whole) may not necessarily be the ultimate goal. Different tasks call for different tools.

7. Consequences of metaphor choice

While I have argued that metaphors can be very useful and legitimate elements of science, considered as both a process and a product, it is not my contention that the use of metaphor in science is never problematic. Because metaphors can function as instruments of manipulation and intervention, they may not only affect how we think of or understand something, they can also be used to alter the nature of things. This is particularly likely in the case of engineering metaphors. Scientists are then frequently not only representing the world but reconstructing it through their choice of language and subsequent efforts to remake the natural world so as to bring it into closer correspondence with the metaphor. Given, then, their power both to mislead and to recreate the things to which we apply them, we need to practice an ethics of attention (Van der Weele 1999) and exercise caution in how we use these metaphorical tools.

The recent discussion regarding whether or not there should be a moratorium on any further research into the "editing" of the genetics of the human germline is an instructive case in point (Lanphier et al. 2015). In their humbler moments, biologists are quick to acknowledge how much they have yet to understand about how organisms develop from fertilized egg to mature adult. It is ironic then that while natural selection, the force actually responsible for concocting this remarkable feat, is regularly described as a "tinkerer," molecular geneticists now talk about whether or not they should use new CRISPR/Cas9 technology to edit the human germline. One might suggest that such language seriously underestimates the difference between editing a literary text or a computer code and manipulating a dynamic genome system with many orders of magnitude greater complexity. Given its potential to mislead as to the degree of precision with which scientists can reasonably hope to predictably alter the outcome, the metaphor of genome editing may be insufficiently accurate for the purposes of discussing the ethical dimensions of this research. Scientists would be better advised perhaps to use the more modest and responsible metaphor of germline tinkering, at least until there is better evidence to support the more optimistic language.[20]

The evolutionary microbiologist Carl Woese (1928–2012) expressed his concern about the consequences of adopting too much of an engineering approach to biology when he wrote: "A society that permits biology to become an engineering discipline, that allows that science to slip into the role of changing the living world without trying to understand it, is a danger to itself" (Woese 2004, 173). Because of the prescriptive nature of metaphor use, that it normalizes certain types of discourse, ways of thinking and acting, we must consider the metaphors we do adopt to talk about and to understand the world, our cells, and ourselves with care and foresight.

Conclusions

Statements of the scientific realism question have tended to conflate two separate issues: Does science aim to provide a *literally* true account of the world? versus Does science aim to provide an *objectively* true account of the world? A *literally* true account would be one in which all the terms and language involved in the account are employed in their standard nonfigurative and nonmetaphorical senses as understood by the relevant community of language speakers. But it seems that when people talk of an *objectively* true account, they often mean one that describes the world as it really is, in its own terms, independently of how it appears to us humans. Obviously this second thesis is much stronger than the first. It seems to me, as it does to many others, that only the first thesis is accessible to us humans or other similarly limited beings. But our desire for *objective knowledge* (i.e., beliefs that are optimally reliable, accurate, and, so far as we can ascertain, true) and for *objectivity in our belief-forming practices* (i.e., using methods of belief-formation that are most likely to result in optimally reliable, accurate, and, as far as we can ascertain, true beliefs) need not commit us to ever achieving the stronger version of the scientific realism thesis. But additionally, I have argued that we shouldn't really be overly committed to the weaker and more attainable version either (a literally true account), for to do so would require jettisoning a lot of very helpful tools and settling for an impoverished version of science. Science is itself a tool or instrument we expect to fulfill several different functions: telling us what the world is like, explaining how or why things happen as they do, and helping us achieve greater control over nature and its events; and for such diverse jobs there may be no one best solution or account. If science can achieve all these things better with the assistance of metaphorical language and thinking, then why should anyone object that metaphors are not literally true?

All our knowledge is knowledge from a human perspective. The expectation that science ought to result in one uniquely correct account is insufficiently founded on a dubious theological-metaphysical thesis that the world was designed and created by a language-speaking agent, to whose privileged "way" the uniquely correct (True) account of science must correspond. Loosening our allegiance to that ontological notion of objectivity and realism does not mean we cannot aspire to objectivity in the methodological sense, as early pragmatists like Peirce (1992) and Dewey argued, and as current philosophers like Ron Giere argues through his perspectivism or perspectival realism.

The focus on science aiming at literally true descriptions of the world, moreover, assumes the crispness of the literal-metaphorical distinction, but like the observable-unobservable and fact-value distinctions, I have suggested that this one too is, in certain cases, less clear and more porous than it may at first appear.

While I have argued in defense of the use of metaphor in science, I have not argued for an indiscriminate use of any or all metaphor. Like other more familiar tools, a metaphor may be appropriate for some specific uses in some particular circumstances and not others. It seems very unlikely that scientists will stop using a set of tools that has proven to be so useful for so long. But this does not mean they cannot be used critically and with careful reflection.

What happens to a metaphor that is deemed inadequate? Four possibilities emerge from this study: (1) The metaphor may stick regardless, e.g., *cell*, but as the scientific understanding of the thing denoted changes, so will the meaning associated with the metaphorical term, and it may become a dead metaphor that has lost all its original connotations; (2) the metaphor may be dropped or exchanged for another nonmetaphorical term, as might have happened to *cell* had Sachs's specially created term *energid* caught on (however this would have involved more than just a change in terminology, it would have resulted in a slightly different, less atomistic understanding of organisms); (3) the metaphor may be dropped or exchanged for another metaphor considered to be more adequate, e.g., the partial shift from talk of *signaling pathways* to *signaling networks*; (4) it may continue to be used, but with a level of awareness that it is only a metaphor and used in an "as if" mode, e.g., kinase *switch*, cell *suicide*. In this last case, the metaphor may not be considered wholly inadequate but adequate up to a point and for a certain purpose.

The other possible career path for a metaphor (*à la* Bowdle and Gentner 2005) is that it not be deemed inadequate at all, but that the term becomes an instance of polysemy. This may be the better explanation of what has occurred with a term like *kinase switch* (or *molecular clutch*): while it is recognized that these proteins are not real *electronic* switches, they may be considered by some scientists to be a different *kind* of switch, a *molecular* switch in a biological cellular system, just as a chair leg is not a real animal limb but another type of leg. Or possibly, as in the case of the expression *cell factory*, what was once a metaphor may, through purposeful manipulation and intervention, become literal. Attempts to modify existing switch-like protein kinases or other molecular genetic components so as to create synthetic *logic gates* and *signaling circuits* may result in a similar turn to literality (see, for instance, Gardner, Cantor, and Collins 2000; Pentimalli 2007; Valk *et al.* 2014).

Finally, I recognize that what I have presented will be considered by some as rather a shaggy dog of an argument. It lacks the clear lines and precision of many other accounts of science, such as hypothetico-deductivism, Popperian falsificationism, Bayesian confirmation theory, or Kuhnian historicism via paradigm-shifts, to name but a few. But philosophers have for too long, in my opinion, attempted to purify science down to a simple and overly refined product (a kind of logical precipitate), when what we ought to be demanding is the natural process itself.

Formulas do have their attractions and their uses, of course, but we need be careful, as scientists often warn one another, not to mistake the cartoon for the reality. Metaphor is a real element of scientific activity, and it is time that it be recognized as such and treated with the seriousness that it deserves. I have been motivated by the question What would our understanding of science look like if we took seriously the reliance placed on metaphor by both the actors (scientists) and by those of us (philosophers, historians, and sociologists of science) who are attempting to compose that theoretical understanding? Whether from sympathy of outlook or from critical disagreement, I hope others will be inspired by this study to move the conversation forward and to improve our understanding of not only the science of cell and molecular biology but of science more generally.

Acknowledgments

I first became interested in the historical development of cell theory in 2003 and gradually my interest shifted to the question, What are all those metaphors in cell biology doing? I have many people to thank for their encouragement and support over the years that I worked on this project. The list includes: Maureen O'Malley, John Dupré, and Staffan Müller-Wille (for the invitation to speak at the "Life of the Cell Workshop" at the Egenis Center, University of Exeter, in April 2009), Kathleen Okruhlik (for inviting me to speak at the "Science, Facts, and Values" workshop in May 2010 at the University of Western Ontario), Lynn Nyhart and Scott Lidgard (for the invitation to participate in the "Biological Individuality and Parts-Wholes" workshop in Philadelphia in May 2012 and in Madison, Wisconsin, in December 2012), Jan Surman, Katalin Stráner, and Peter Haslinger (for inviting me to talk at the "Nomadic Concepts: Biological Concepts and Their Careers beyond Biology" workshop at the Herder Institute, Marburg, Germany, in October 2012), Jane Maienschein, Manfred Laubichler, and Karl Matlin (for the invitation to be part of the "Updating Cowdry's *General Cytology*" workshop at the Marine Biological Laboratory in Woods Hole in October 2014). Lorraine Daston also arranged for me to give a talk on an earlier version of what became chapter 3 during a sabbatical visit at the

Max Planck Institute for History of Science in Berlin-Dahlem in the fall of 2011. The opportunity to attend these events and to learn from the organizers and the other participants is truly appreciated. Gordon McOuat, Evelyn Fox Keller, and John Tyler Bonner were all generous with their time and advice, especially in the earlier stages of the project. Scott Gilbert, Brian Hall, Carlos Sonnenschein, and Ana Soto were all exceptionally accommodating, kind, and helpful and allowed me to visit with them in their labs to talk about metaphor in science. I thank the librarians and staff at Cape Breton University for providing assistance with various interlibrary loans and literature searches. Ariane Dröscher, Richard Keshen, and Andy Parnaby read drafts of chapters and provided enormously helpful feedback. Other friends and colleagues too many to mention (from philosophers, historians, sociologists, to biologists) have at conferences and workshops listened to my talks and challenged me (in Q & A sessions or over drinks) to improve and to expand my ideas on the topic —Thank you all. I am also grateful for the helpful comments of two anonymous reviewers for the University of Chicago Press, to Louise Kertesz for her expert copyediting and for spotting the many omissions and errors in my initial bibliography, and to Karen Merikangas Darling, the executive editor, for taking the project on. Of course, I alone am responsible for any of the arguments and blunders remaining in the final product. Christie MacNeil, digital archivist at the Beaton Institute at Cape Breton University (and a graduate of the masters program in History and Philosophy of Science and Technology at the University of Toronto), has been invaluable in preparing the illustrations, ensuring I had all the permissions required to reproduce them, and keeping all the files properly organized. Support for some of this research, especially earlier on, was provided by a Social Sciences and Humanities Research Council of Canada Grant (2005–2008), and by several research grants and teaching releases from Cape Breton University. Last, but most importantly, I must thank my family: my wife Kellie and daughters Clara and Alice for all their love and support.

I have discussed some of the issues covered in this book in previous publications. Chapter 1 draws on some material previously published in "The Theory of the Cell State and the Question of Cell Autonomy in Nineteenth and Early Twentieth-Century Biology," *Science in Context* 2007, 20(1): 71–95; "Amoebae as Exemplary Cells: The Protean Nature of an Elementary Organism," *Journal of the History of Biology* 2008, 41: 307–37, with permission of Springer; "Ernst Haeckel and the Theory of the Cell State: Remarks on the History of a Bio-Political Metaphor," *History of Science* 2008, 46(2): 123–52; and "The Redoubtable

Cell," *Studies in History and Philosophy of the Biological and Biomedical Sciences*, special issue, "Historical and Philosophical Perspectives on Cell Biology," 2010, 41(3): 194–201, with permission from Elsevier.

Chapter 2 contains some material previously published in "The Cell's Journey: From Metaphorical to Literal Factory," *Endeavour* 2007, 31(2): 65–70, with permission from Elsevier; and "In Search of Cell Architecture: General Cytology and Early Twentieth-Century Conceptions of Cell Organization," in *Visions of Cell Biology: Reflections Inspired by Cowdry's "General Cytology,"* edited by Karl Matlin, Jane Maienschein, and Manfred Laubichler, University of Chicago Press, 2018.

Chapter 3 draws on some material previously published in "The Deaths of a Cell: How Language and Metaphor Influence the Science of Cell Death," *Studies in the History and Philosophy of Biological and Biomedical Sciences* 2014, 48 (B): 175–84, with permission from Elsevier; and "Discovering the Ties That Bind: Cell-Cell Communication and the Development of Cell Sociology," in *Biological Individuality: Integrating Scientific, Philosophical, and Historical Perspectives*, edited by Scott Lidgard and Lynn K. Nyhart, 109–28. University of Chicago Press, 2017.

Chapter 4 also contains some material previously published in "Discovering the Ties That Bind: Cell-Cell Communication and the Development of Cell Sociology," in *Biological Individuality: Integrating Scientific, Philosophical, and Historical Perspectives*, edited by Scott Lidgard and Lynn K. Nyhart, 109–28, University of Chicago Press, 2017.

Nearly all of this previously published material has been rewritten for the purposes of this book.

Notes

INTRODUCTION

1. I agree with Hannah Landecker's remark that "the rise of cell signalling is, I think, as important a development in the history of life science as was the original cell theory" (Landecker 2016, 90).

2. Technically the book should be called *The Fourth Lens* because the human eye contains a lens of its own through which the image must first pass before it is processed through the figurative lens of our conceptual system and its metaphorical filters—(and Wilson is wearing glasses that introduce even a *fifth* lens!). I hope the simplification will be excused for the sake of a snappier title.

CHAPTER I

1. *Oxford English Encyclopedia*, online edition. Accessed October 28, 2016.

2. See Kay (2000), Keller (1995, 2000, 2002) for the history of how these information and computer metaphors became so entrenched in molecular genetics and biology.

3. For more on the relationship between metaphors and models, see Bailer-Jones (2002) or (2009).

4. See Hall (1969, 179–92) for a more extensive discussion of this issue.

5. Leeuwenhoek seems not to have used the term "cell" at all in his letters to the Royal Society in which he described his observations with his microscopes. This may be because, as Dobell (1960, 44) explains, Leeuwenhoek spoke and understood only Dutch. He did though refer to some of his animal-

cules as "living Atoms" (or perhaps this was an innovation on the part of the letter's translator into English), see Dobell (1960, Plate XVIII, facing p. 113), and he also used the term "globule" when speaking of what were evidently red blood cells (Harris 1999, 16.).

6. See for instance Canguilhem (2008), Duchesneau (1987), Jacyna (1984), Jacyna (1990), Harris (1999), Parnes (2000), Nicholson (2010).

7. Attributions of priority for any scientific development are of course wrought with difficulty (see Kuhn 2012, 55ff.), and there are many other individuals deserving of recognition for their observations and intellectual theorizing about cells. But discussion of these matters is outside my main concern here. Those interested in a more complete history of the early developments of the cell theory will find Baker (1948–52), Hughes (1959), Hall (1969), Duchesneau (1987), Harris (1999), and Dröscher (2014a) useful.

8. As Nyhart and Lidgard (2011) explain, the question of delineating the parts of an organism from the organism as a whole was a matter of central concern throughout nineteenth century biology, and it continues to be so today.

9. De Bary is frequently quoted as having said this, but the only line at all similar to this that I have been able to find appears in a book review in which he laments the "Hegemonie der Zelle," established by Schleiden, which has led to the popular conviction that "die Zelle die Pflanze und nicht umgekehrt die Pflanze Zellen bilde." See de Bary (1879).

10. There are exceptions to this rule among the "lower" invertebrates, demonstrated most famously by Abraham Trembley's (1710–84) experiments with freshwater polyps.

11. See Müller-Wille (2010) for an account of the merger of cell theory and the theory of inheritance.

12. Similar comparisons had been made between various organs of the human body and social organizations long before the advent of the cell theory. See Koschorke et al. (2007).

13. The idea of a physiological division of labor was introduced by Henri Milne-Edwards (1800–85) (1851). For the history of the transfer of the division of labor principle from political economy to biology, see Young (1990), Limoges (1994), and D'Hombres (2012).

14. Virchow (1855, 25), quoted in Temkin (1949), 175. See Ackerknecht (1953); Mazzolini (1988); Otis (1999); Goschler (2002), and Johach (2008) for extensive discussion of Virchow's social and political metaphors.

15. On Georg Bronn's translation of Darwin's ideas into German and the challenges it posed, see Gliboff (2008).

16. Translations are my own unless indicated otherwise.

17. Haeckel included bacteria in the Monera, but the forms he considered most primitive and most likely to represent the Ur-organism from which all life evolved were amoeboid in nature. See Reynolds (2008a).

18. See Reynolds and Hülsmann (2008) for more detailed discussion of Haeckel's theory of early metazoan evolution.

19. Haeckel called these simple cell communities "coenobia" (after the term used to designate a monastic community), derived from the Greek "koi-

nos," meaning "common" (*Concise Oxford English Dictionary*). Perhaps he was making an explicit link to the original cell metaphor.

20. This list omits a couple of levels ("antimeres" and "metameres") having to do with multicellular morphological structures found in plants and animals.

21. For more details of Haeckel's use of the cell-state metaphor, see Reynolds (2008b).

22. See Nyhart and Lidgard (2011) for discussion of the question of biological parts and wholes with special attention to the Siphonophora.

23. See Richards (2008) 185–89 for more discussion of these experiments and their significance for Haeckel's theorizing about embryology and phylogeny, etc.

24. See, for instance, Cleland (1873).

25. Mendelsohn (2003) provides an insightful account of the history of the cell theory through the diverse sorts of materials that were considered to be exemplary of what a cell was supposed to be.

26. Geison (1969). Not every supporter of the protoplasm concept drew from it mechanistic conclusions. Lionel Beale (1828–1906), for instance, who divided protoplasm into a living "Bioplasm" substance and a nonliving "formed matter" created by the former, was an avid vitalist and critic of Huxley's physicalism. See Strick (2000), chaps. 2, 4, 5.

27. I thank Ariane Dröscher for emphasizing this point to me.

28. See Aschoff, Küster, and Schmidt (1938), 193ff., for response to Heidenhain.

29. Haeckel (1866), I: 528, "Die plastiden oder Plasmastücke bilden als die morphologischen Individuen erster Ordnung die Bausteine, aus deren Aggregation sich der Körper aller Organismen aufbaut . . ."; also in *Nat. Sch.* 1st ed. (1868), 285; 2nd , 307 and later.

30. The surprising result of tissue and cell culture techniques, revealing that animal cells can indeed remain alive separated from the body, is treated in chapter 3.

31. The cytoplasm of chondrocytes does have a filamentous network appearance while they are actively producing the collagen and proteoglycan proteins of which hyaline cartilage is composed.

32. Original in Drysdale (1874), 104.

33. The internal complexity of such microscopic infusoria was the basis for Christian Gottfried Ehrenberg's claim that they were "complete" animals possessing a digestive system with all the necessary internal organs (Ehrenberg 1838).

34. The lament made by both Whitman and Dobell points to an example of the theory-ladenness of observation thesis. In fact, Norwood Russell Hanson opens his discussion of this topic with an allusion to Dobell's complaint about how two microbiologists looking through a microscope at the same organism can see two different objects: one a unicellular animal, the other a noncelled animal (Hanson 1958, 4).

35. Huxley (1853, 265–66) had also raised this objection. Though neither used this analogy, they might have said that this should make as little sense as calling a brick or building stone a complete house.

36. See Landecker (2007) for a history of these developments and their more general implications for biology and biotechnology.

37. Morange (2000), Bechtel (2006), and Moberg (2012) are only a few very useful resources for the history of molecular biology. Reynolds (2018) discusses early twentieth-century attempts to discern the cell's internal organization.

38. Reynolds (2010) provides a lengthier treatment of criticism of the cell theory from the nineteenth century to the present.

39. Similar statements appear in the second edition (1900, 17) and the third (1925, 4).

40. These embryological experiments illustrate two key developmental patterns, known as regulative and mosaic. Haeckel's successful experiments in artificially cleaving siphonophore embryos dealt with regulative development, as did the more famous experiments performed in 1892 by his former student Driesch, which showed that isolated sea urchin blastomeres continue to grow into complete larvae, though of diminished size. Cells in these cases are said to be conditionally specified, meaning that the specific type of tissue they result in is reliant on their location within the developing embryo, so that if a piece of the embryo is removed, the remaining cells become respecified to accommodate or regulate for the disruption. The second type of experiment wherein cells continue to behave as if they have not been displaced is an example of mosaic development. In this case the cells are said to display autonomous specification and will continue to develop their specific tissues independent of their location, as if behaving automatically or autonomously of the rest of the organism. Embryos developing in this fashion appear as a patchwork or mosaic of separate bits. This behavior was first documented in 1887 by Laurent Chabry (1855–94), but made better known by Wilhelm Roux (1850–1924), a student of Haeckel, who in 1888 carried out defect experiments on two- and four-celled frog embryos by killing one or two blastocysts and obtaining half embryos (Gilbert 2003, 61).

41. See Maienschein (1991b) and Dröscher (2008) for discussion of Wilson's choice of illustrations and diagrams.

42. See especially pp. 136–38.

43. See Stanier and Van Neil (1962) for discussion of the accommodation of bacteria into the cell theory.

44. As just one instructive example, see Puck (1972).

45. See Meinesz (2008), chapter 7, "The Lego Game."

46. Clément (2007) discusses the pedagogical problems created by the common visual presentation of animal cells in textbooks as isolated cells, leaving the viewer potentially puzzled as to how or why cell differentiation takes place.

CHAPTER 2

1. Reynolds (2018) discusses an example of the cells are batteries metaphor from the early twentieth century.

2. Nelkin (1994) has also explicated what she calls a promotional function

of metaphors, which is a specific type of rhetorical usage of figurative language to gain support for a particular hypothesis or project.

3. One could add to this a scientist's personal notebooks or laboratory notebooks.

4. For the notion of "theory constitutive" metaphors, see Boyd (1993).

5. Quoted in Harris (1999, 32). The original is Raspail (1843) I, 28.

6. The French term for factory is "usine," "manufacture," or "fabrique," but as Maxine Berg (1985, 41–42) explains, there was a continuous development in the late eighteenth to nineteenth centuries from workshop to factory systems of labor organization. Milne-Edwards may also have chosen "atelier" to emphasize the division of labor among the various organs within the living body, which itself would be analogous to the factory as a whole. I am grateful to Mary Morgan for drawing my attention to the relevance of the history of the factory system.

7. Kohler (1982, 348, n. 81) mistakenly lists the title as "The laboratory of the living cell," which is understandable given the talk's focus on the cell.

8. Kyne and Crowley (2016) provide a discussion of these contrasting viewpoints from both historical and current perspectives.

9. Quoted in Landecker (2007), 191.

10. See Matlin (2002) on how molecular cell biologists learned to decompose the cell into smaller components while retaining essential structure and organization.

11. Quoted in Bechtel (2006), 118.

12. http://ec.europa.eu/research/fp5/eag-cell2.html. (Accessed 25/04/2014.)

13. Among the list of project objectives are included: "To transform the tomato fruit into a cell factory for carotenoids: overproduction of lycopene, zeaxanthin, astaxanthin and lutein"; and "To transform the potato tuber into a cell factory for carotenoids: overproduction of lycopene and beta-carotene." http://europa.eu.int/comm/research/quality-of-life/cell-factory/volume1/projects/qlk3-2000-00809_en.html. (Accessed 09/14/2006.)

14. The metaphorical term "enzyme equipment," common among biochemists, can be traced back as far as Shull (1922).

15. See also Rietman, Colt, and Tuszynski (2011) for further example of how industrial factories are being used as models for more efficient projects in synthetic biology.

16. See Feest (2010) on the idea of concepts as tools.

17. I skip over here the metaphors of molecular "postal" and "zip codes," which scientists use to account for how proteins and other macromolecules reach their appropriate "addresses" in the busy context of intracellular "traffic."

18. http://sciencenetlinks.com/lessons/cells-2-the-cell-as-a-system/ and http://sciencenetlinks.com/student-teacher-sheets/comparing-cell-factory-answer-key/. (Accessed 20/04/2014.)

19. Lorraine Daston initially asked me when I presented a version of this chapter at the Max Planck Institute for History of Science in 2013 why no one thought of protein synthesis in terms of a farming or agricultural metaphor. The possibility of a kibbutz metaphor with its particular social organization

highlights even more the contingency of the factory metaphor and the alternatives left unexplored.

20. Wikipedia, "Cellular Manufacturing." Accessed 14/03/2014.

21. See Coleman (1977), especially chapter 6, and Allen (1978), chapter 4, for good overviews. Nicholson (2013) provides a more recent critical analysis of the "machine conception of the organism."

22. See also Rather's introductory remarks (21–22) to Virchow (1958).

23. Such was the view of Huxley (1868) as expressed in his influential lecture on protoplasm as the physical basis of life.

24. http://biobricks.org (accessed 3 June 2014). It should be noted that the BioBrick Foundation is a public-benefit organization founded on the principle of free and open access to the technologies and knowledge created.

25. http://igem.org (accessed 4 June 2014). See Roosth (2017) for a very interesting discussion of both projects.

26. See Dutfield (2012) for discussion of the metaphors of synthetic biology and intellectual property law.

27. See Pigliucci and Boudry (2011) and Boudry and Pigliucci (2013). During a visit with Michael Behe in 2011, I asked about this, and he cited the paper by Bruce Alberts (1998) as evidence for the claim that cells are indeed machines, which remain black boxes, inexplicable by the principles of Darwinian evolution.

28. Lewontin (1991) made similar criticisms.

CHAPTER 3

1. See Landecker (2007) and Skloot (2010) for the story of the HeLa cell line, derived from a cervical tumor of a young woman in 1951 and still living today in laboratories the world over.

2. Carrel explained again later in his popular book *Man the Unknown* that cell sociology refers to the physiology and anatomy of tissues and organs (Carrel 1937, 77).

3. Cited in Dunn and Jones (1998), 124.

4. This is the case not only for cells from a clonal cell culture derived from a single cell, but for cells taken from a mixed population as well. Cell proliferation is a mass effect requiring a minimum number of cells, or at least a minimal concentration of cell products, e.g., growth factors.

5. Fischer rejected, however, the interpretation that a tissue culture represented a colony of cells. In his opinion, tissue cultures exhibited a rudimentary form of organization characteristic of tissue structure in vivo, and for that reason should be regarded as tissues, not colonies of autonomous elementary cell-organisms (Fischer 1923).

6. Cancerous cells are, as it were, deaf to the messages of their fellow tissue cells telling them to maintain an orderly existence. This is due in part at least, as Loewenstein and Kanno (1966) discovered, to their failure to form gap junctions with other cells through which communication can take place. In this sense, communication literally means contact. Is it a testament to this ancient form of communication that we still commonly use phrases like "I'd like to get in contact with so-and-so" or "Please put me in touch with X"?

7. Of course, those opposed to the "cell standpoint" and who refused to regard the organism as an aggregate of distinct, independent cells had little doubt that there was a free and constant communication of material and organizing influence throughout the continuous protoplasmic mass of the organism.

8. As one recent commentator puts it: "The metaphor of cell-to-cell communication implicitly elevates the cells to the status of self-motivated individuals each possessing some knowledge and communicating with each other in the business of maintaining the tissue architecture" (Rosenfeld 2013, 227).

9. See Wolpert (2008) for a general account. Keller (2002, 173–97) situates Wolpert's theory in historical and philosophical context.

10. Although he does not mention her by name, Wolpert seems essentially to have conceded the sort of criticisms Chandebois voiced of the morphogenetic field concept and has adopted a position strikingly similar to the cell sociology perspective (cf. Wolpert 2009, 92ff. and Wolpert 2008 [1991]).

11. More generally, cell biologists call a cell phenotype or trait "cell autonomous" if it is not dependent on the presence of other (possibly mutant) cells or cell types.

12. Though Chandebois does not use this particular image, her point might be put this way: the cells of an embryo are not Leibnizian monads, each running on their own autonomous but parallel programs. This would make the coordinated development of a complete and integrated organism the result of a miraculous preestablished harmony.

13. Chandebois tends to use metaphors and analogies of a civilization or society in development because she is interested in the development of an animal as a whole, though she does at times make the analogy between an individual cell's development and personal developmental psychology.

14. One might think of the acquisition of an elementary social behavior as akin to speaking a particular language, which does not fully determine the type of person one ultimately becomes but does impose certain proclivities and restrictions on the interactions one has with individuals of different groups.

15. Her later writings suggest there may even be a hint of more traditional vitalism in her opinions. Chandebois's later publications have been devoted to critiques of gene-centric and chance-driven "neo-Darwinian" accounts of evolution and of the moral status of the unborn human embryo. These writings have been well-received by the political right, principally in France, while a lack of English translations prevents them from getting much notice in North America. Intelligent Design websites occasionally do contain translations of brief summaries of her views. She appears to favor what might be described as an epigenetic-orthogenetic theory of evolution premised on a tight link between ontogeny and phylogeny (Chandebois and Faber 1983, 180–83).

16. There is in fact at least one "Laboratory of Cell Sociology" at the National Institute for Basic Biology, in Japan. (http://www.nibb.ac.jp/en/sections/cell_biology/hamada/index.html.)

17. The expression "programmed cell death" was introduced by the developmental physiologist Richard Lockshin. See Lockshin and Williams (1964). Programming and other computer engineering metaphors will be treated in

chapter 4. For a more detailed discussion of the history of the various meta-phors informing the science of cell death, see Reynolds (2014).

18. Keller (2002) also argues that metaphor has been central to the articulation of explanations in developmental genetics.

19. Levine and Davidson's (2005) introduction to the special issue of *PNAS* devoted to gene regulatory networks makes strong causal and explanatory claims couched in these metaphorical terms.

20. Bechtel (2010) tells a story of how decomposition of the cell into its parts (cell as locus of inquiry) then requires recomposition to understand how the cell functions as an integrated autonomous whole/system (cell as object of inquiry). Cell sociology à la Chandebois is not exactly about recomposition following decomposition, but an appreciation of the phenomenology of the cell as a social organism and its group behavior, akin to Abercrombie's ethological or natural history approach to the cell.

21. Of course, atoms, stones, and machines also have histories and stories about how they came to be, but they are not so relevant to understanding their current properties and behavior.

CHAPTER 4

1. The conscription metaphor no doubt had strong resonance in this period between the two world wars. I am grateful to Jamie Elwick for bringing Keith's essay to my attention.

2. Paul Ehrlich (1854–1915) introduced the receptor concept in 1900 to explain the specific affinity of chemically active agents like toxins for particular target cells. Ehrlich used the metaphor proposed earlier by the chemist Emil Fischer in 1894 that an enzyme and its substrate fit like a "lock and key" to explain the specific affinity between a toxin and its receptor. The physiologist John Newport Langley (1852–1925) then proposed in 1905 that "receptive substances" on the cell protoplasm explained the specific action of other agents, such as drugs and the hormones recently discovered by Starling and Bayliss. The lock and key metaphor has remained an important model of ligand-receptor activity throughout the twentieth century to the present. See Cramer (1994), Maehle (2009).

3. G proteins, or guanosine nucleotide-binding proteins, are a family of proteins that act as molecular "switches" in the activation and deactivation of various intracellular signaling pathways.

4. Norton Zinder (1928–2012) later recalled that "we were convinced we had a reasonable explanation for the Salmonella phenomenon [i.e. the transference of biochemical markers between strains of bacteria separated by a filter small enough to allow virus to pass through but not bacteria] and Lederberg suggested that we call it 'transduction.' Other words such as 'entrainment' were considered and wisely rejected" (Zinder 1992, 293). Lily Kay writes that Joshua Lederberg (1925–2008), the pioneering molecular geneticist who shared the Nobel Prize in physiology or medicine with Edward Tatum and George Beadle in 1958, was initially critical of the use of information discourse in molecular genetics, but he too eventually adopted the practice: "A new way of thinking and speaking began to permeate molecular genetics.

Living entities were increasingly conceptualized as programmed communications systems, in which, as Lederberg astutely sensed, instructions and material content were collapsed into a single amorphous fabric of information" (Kay 2000, 114).

5. Kay (2000, 23ff.) makes a similar point about the use of terms such as signal, message, and information more generally in the application of cybernetic theory to molecular biology.

6. Fibroblast growth factor (FGF), for instance, in the early stages of embryogenesis induces the formation of mesoderm, while later induces the creation of the nervous system. Bone morphogenetic protein (BMP) in the presence of the Wnt (wingless) signal means "build cartilage," but in its absence leads to cell suicide (Niehoff 2005, 111, 132). As the cell biologist Guenter Albrecht-Buehler cautions: "Cell biological information is a context-dependent quantity. The more we decompose it into its molecular letters the more we destroy its meaning, which ultimately contains the profound explanation we seek" (Albrecht-Buehler 1990, 192).

7. According to Sir Paul Nurse, cell biologist and 2001 Nobel Prize Laureate for work on cell cycle regulation, the cell is "chemistry made into biology." Quoted in Landecker (2007, 5).

8. Scientists have been thinking of cells in electrical terms for a while, of course. As a matter of fact, all electrical charges in the cell are carried by chemical ions and so, as the molecular biologist Dennis Bray explains, the distinction between mechanics and chemistry is a human creation of no real importance to the cell (Bray 2009, 93).

9. See, for instance, Tepperman and Tepperman (1965). Of course, another source of a switch metaphor in biology derives from the work of François Jacob and Jacques Monod's operon model of gene regulation in the late 1950s and 1960s. See Jacob (1979), Kay (2000) and Garcia and Suárez (2010).

10. For a sampling of some of the earliest instances of the signaling network metaphor, see Leclercq and Dumont (1983); Sugimoto et al. (1988); Sternberg and Horovitz (1989); Bray (1990); Forgacs (1995); Pawson (1995); Pawson and Saxton (1999); Bhalla and Iyengar (1999).

11. This is the most highly cited paper ever published in the journal, with 24,442 citations as of Dec. 30, 2016, according to Google.

12. As just one example illustrating how this metaphor has caught on, consider the following review article title: "Wiring the Cell Signaling Circuitry by the NF-κB and JNK1 Crosstalk and Its Applications in Human Diseases" by Liu and Lin (2007).

13. See, for instance, Bird (2002) and Xiong and Ferrell (2003).

14. Proteins are said to have four levels of structural organization. *Primary* structure refers to the linear sequence of amino acids; the *secondary* structure refers to the α helices and β sheets formed by stretches of polypeptide chains; *tertiary* structure refers to the three-dimensional folded shape of the polypeptide chain. When a protein consists of a complex of more than one polypeptide chain, biologists refer to this as the *quaternary* structure (Alberts et al., 2008, 136).

15. For these reasons, Pappas (2005) recommends replacing the blueprint

metaphor for the genome and proteome with a theatrical cast of characters (*dramatis personae*) metaphor.

16. Silvia Svegliati et al. 2014.

17. Ronen Hope et al. 2014.

18. There is another related tradition of speaking of cells and their protein components in the metaphorical terms of ecology; see, for instance, Welch (1987), Fisher, Paton, and Matsuno (1999), and Nathan (2014). The concept of molecular ecology was introduced by the embryologist and developmental biologist Paul Weiss (cf. Weiss 1947). This is another instance of metaphor use in cell biology worthy of greater attention, but one I must leave aside for the moment.

19. See, for instance, van Roey, Gibson, and Davey (2012) "Motif Switches: Decision-Making in Cell Regulation."

20. See Boniolo (2013) and Blasimme, Maugeri, and Germain (2013) for further refinements of the notion of a mechanistic explanation in biology in light of the complexity of cell signaling pathways.

21. For an interesting discussion of what a mechanistic explanation might look like from a nonreductionist and feminist perspective, see Fehr (2004).

22. See Pawson and Scott (1997) for a review. Pawson worked out how the highly conserved SH2 (*Src homology* 2) domain allows some proteins to bind other signaling proteins and receptors together through phosphotyrosines within specific peptide sequences (motifs), thus acting like an electrical adaptor plug to bring several components of a signaling pathway together. Some protein domains possess specific three-dimensional structure, which allows them to bind and interact with corresponding peptide motifs within the same or separate proteins. They may have catalytic function or, as is the case with the interaction domains, promote the interaction of proteins and other cellular components and the formation of signaling complexes. As they appear as repeated motifs across widely different taxa—from viruses to humans—they constitute another level of biological and evolutionary modularity.

23. See, for instance, Guilluy, Garcia-Mata, and Burridge (2011); Wang (2011).

24. http://www.syntheticbiology.org. Accessed 28 August 2014. See O'Malley et al. (2007) and Keller (2009a; 2009b) for more critical discussion of the meaning(s) of synthetic biology.

25. The team led by Celera Genomics founder Craig Venter that "booted-up" a synthetic genome into a bacterial cell in 2010 coded into it a version of the Feynman quote—a misquote, as it turns out, "What I cannot build I cannot understand," according to a story in the *New Scientist*.
www.the-scientist.com/?articles.view/articleNo/29636/title/News-in-a-nutshell/. Accessed Sept. 3, 2014. As a relevant aside, synthetic biologists refer to the living cell minus the genetic program as a "chassis," in analogy to the structural framework minus the engine in automobile manufacturing or the framework minus the circuit board in electronics manufacturing. See Danchin (2012).

26. Though it should be noted that many in the synthetic biology community, e.g., the International Genetically Engineered Machine (iGEM) Founda-

tion, which supports the Registry of Standard Biological Parts, are in favor of keeping the results as "open source," freely available to anyone who wishes to use them, within responsible and ethical boundaries. See Roosth (2017).

27. On the ubiquity of crosstalk between signaling pathways in humans, see Korcsmáros et al. (2010). See Rajasethupathy, Vaytadden, and Bhall (2005) for discussion of the Vioxx and crosstalk connection.

28. Efforts in this direction are underway by multiple teams of researchers, but a good one-stop site to explore is VCell The Virtual Cell website of the National Resource for Cell Analysis & Modeling at the University of Connecticut Health Center: http://vcell.org. A landmark achievement is the construction of a whole-cell computer model of an entire unicellular bacterium, the human pathogen *Mycoplasma genitalium*, including all of its molecular components and their interactions, by a team at Stanford (Karr et al. 2012). This is the same organism into which the team at the J. Craig Venter Institute "uploaded" their synthetic genome.

29. Sismondo (1996, 143–44) makes a similar point that metaphor use in science is a "performative speech act."

CHAPTER 5

1. In addition to the research alluded to earlier on code and program metaphors in molecular genetics, another of more recent significance is the metaphor of "stem" cells, about which see Maehle (2011), Brandt (2012), and Dröscher (2012; 2014b).

2. On the naturalistic approach in philosophy of science, see Callebaut (1993).

3. Black spoke of a metaphor as consisting of a *focus* and a *frame,* but the terms "source" and "target" are more intuitive and have become more standard. The target of the metaphor is the subject to which the metaphor is being applied, and the source is the term or concept from which the metaphorical description is being drawn.

4. Recall the discussion of the cell factory metaphor in chapter 2.

5. An earlier suggestion as to the connection between metaphors and models in science was made in Black (1962b).

6. Note the interesting mix of metaphors here——of eye and hand, of seeing and grasping—to which we will turn shortly.

7. See Pinker (2007), 253ff. for an approving discussion.

8. Runke's account is developed more fully and in greater detail in Runke (2008). I am grateful to her for sharing her PhD thesis with me.

9. Grant's original example, taken from Camp—who was following Boyd (1993)—involved the metaphor that memory storage and retrieval is the opening of a computer file; I have substituted my own example.

10. Davidson (1984, 262) does compare the effect of metaphor to a bump on the head, insofar as both may lead us to notice something.

11. Hesse (1966) further explained that any scientific metaphor will establish three types of analogies between the source and target systems: positive, negative, and neutral. A productive metaphor will generally display more positive than negative analogies, and its invitation to explore the neutral analogies

(those about which it is still uncertain whether they are positive or negative) is a major reason why metaphor is of such service to science.

12. This topic was given extensive treatment in Rorty (1979).

13. For example, Richards (1955), Black (1962b; 1977), Miller (1996), Van Rijn-van Tongeren (1997), Bradie (1998), Brown (2003), Ruse (2005), Camp (2006), Haenseler (2009), Rodriguez and Arroyo-Santos (2011). Feest (2010) talks of concepts in general as tools analogous in important respects to physical tools and instruments.

14. See Bailer-Jones (2002) and Contessa (2011) for general overviews of the relation between models and metaphors.

15. Lakoff and Johnson say as much themselves: "So far as metaphor is concerned, the Neural Theory of Language attempts to explain on neural grounds why we have the primary metaphors we have" (Lakoff and Johnson 2003, 264).

16. This is probably unfair to attempts like that of Steinhart (2001), who doesn't exactly attempt to give a literal account of how metaphors work but one invoking the tools of formal logic. His Structural Theory of Metaphor sets out to give a precise account of the "cognitive meaning" of metaphors, but insofar as it is a "structural" theory appealing to the "logical space" of possible worlds, it too seems to draw rather heavily on metaphor to make its case.

17. An alternative description is "conventional metaphor," as contrasted with "novel metaphor," but some, e.g., Lakoff (1987), insist that conventional and dead metaphors are distinct sorts not to be confused or conflated.

18. Griffiths (2001) calls the concept of genetic information a "metaphor in search of a theory."

19. Some readers might have concerns that the language used in an article summary is not representative of the "real" content in the body of the paper, but a casual read through any scientific article should reveal the language used in the summary is generally equivalent to that used in the introduction, methods, results, and discussion sections of the paper. Titles, on the other hand, may engage in more playful or attention-grabbing metaphor. This is especially true of reviews and shorter correspondence pieces.

20. Recall from chapter 4 that kinases are often characterized as switches that turn on or off other functional proteins by transferring to or from them a high-energy phosphate group.

21. A blog describing the article refers to $\mathfrak{r}$CaMKII as the "wagon" for the Ca^{2+} signal. Neuroscience Institute, New York University. 2014. "$\mathfrak{r}$CaMKII: The wagon for Calcium-Calmodulin complex." http://neuroscience.med.nyu.edu/node/786. Accessed Jan. 16, 2015.

22. Van Fraassen argues further that explanations are always responses to more particular why questions situated against a background of contrast classes, e.g., "Why did X happen rather than Y or Z?"

23. Just such discussions were the subject of several talks at a workshop on the past, present, and future of cell biology organized by Jane Maienschein and Karl Matlin at the Marine Biological Laboratory in Woods Hole in the

fall of 2014. Some of the papers from this meeting will be published as Matlin, Maienschein, and Laubichler (2018).

24. See, for instance, Hempel (1965, 240); Salmon (1989); Trout (2002). In defense of understanding as a goal of explanation, see De Regt (2009) and Grimm (2010); and for a critical response, Khalifa (2012).

25. I do not use these terms in their typical evolutionary senses, wherein a trait shared by two organisms is said to be homologous just in case they both possess it as a result of a shared ancestry. "Homologous" is used here in its more general and imprecise sense (as defined by the *Oxford English Dictionary*) of having a "sameness of relation," to which I would add, a greater than superficial similarity or mere analogy.

26. This quotation is typically traced back to Lewontin (2001) who himself gives no reference. A close and earlier anticipation of this phrase can be found in Braithwaite (1960, 93): "The price of the employment of models is eternal vigilance."

27. In a similar way Baetu (2012b) defends genomic program analogies by arguing that they should not be construed as purportedly complete and exhaustive descriptions of cellular physiology, but rather abstract mechanism schemas serving specific pragmatic research objectives.

28. See, too, Stepan (1986, 272), who says of eighteenth- and nineteenth-century analogies between race and gender, "The metaphor, in short served as a program of research."

CHAPTER 6

1. As an empiricist, van Fraassen maintains that all we ought to be committed to is the claim that "science aims to give us theories which are empirically adequate" (Van Fraassen 1980, 12), by which he means that what a theory says of the observable things and events within its scope are true. Empiricists consider the attempt to penetrate beyond the observable phenomena a misguided foray into metaphysics and that science should stick with "saving the phenomena."

2. Maxwell (2013[1962]) and Hacking (1983, chapter 11) provide arguments in favor of saying we do observe things with a microscope. Van Fraassen (2008, 93–113), on the other hand, disagrees. Microscopes, he says, create the very "images" that we are able to "detect" by their use. These images, whether created by optical or more technologically sophisticated instruments like scanning electron microscopes, are, according to him, a type of artifact. Microscopes are "engines of creation" rather than "windows upon an invisible world" (Van Fraassen 2008, 101), and so we should not claim to have direct observable knowledge of the entities they purport to show us. Hacking argues that the chief reason we become convinced of the reality of those things we investigate with the aid of even a light microscope is that we can interfere with and manipulate them in quite reliable ways. In short, we trust what we see with a microscope not because of any theory about the instrument or the entities seen with it, but for the pragmatic reason that "we learn to move around in the microscopic world" (Hacking 1983, 209).

3. Anjan Chakravarrty (2014), however, seems to disagree about this.

4. See Hacking (1999) for a thorough and helpful critical discussion. Brown (2001) provides a valuable discussion of the broader issues involved in the Science Wars.

5. Slater (2013) considers the question whether distinct types of differentiated cells are natural kinds but does not address whether the concept of cell itself is one.

6. Consider this: if biologists decided tomorrow that viruses count as live organisms, would viruses then also count as cells, or would the claim that all living things consist of one or more cells be refuted?

7. This was the title of a talk Cavalier-Smith gave at the "Life of the Cell" workshop held at the Egenis Center at the University of Exeter in 2009: "The Evolution of Cells: Real History Is Messy and Non-Platonic."

8. Brown's embodied realism appears to be equivalent to the position Lakoff and Johnson call "experientialism" (Lakoff and Johnson 2003, 226–228).

9. Hacking (1999) refers to these sorts of questions as "sticking point #2" between the realists or "inherent structurists" and the constructionists in the Science Wars. Searle would count as a strong advocate of the world having its own inherent structure independent of us, while those Hacking describes as "nominalists" (with whom he expresses strong sympathy) are less convinced of this (Hacking 1999, 83).

10. I do not pretend that my account of the psychology behind this etymology is anything more than a "just so story." I offer it only as a plausible interpretation. See Daston and Galison (2007, 29–31) for a more complete account of the surprisingly twisted history of these terms' meanings.

11. In 2015, the Supreme Court of Canada ruled against the practice of beginning municipal council meetings with a public prayer. The particular prayer that had been used in my own municipality began: "God our creator, bless us as we gather today for this meeting; You know our most intimate thoughts; Guide our minds and hearts so that we will work for the good of the community, and help all your people."

12. Richard Rorty also argued at length for substituting the metaphysical conception of objectivity with the social ideal of solidarity and intersubjectivity, beginning with Rorty (1979).

13. Although many of my undergraduate students frequently tell me that "everyone sees the world differently," —which, if true, would make our ability to communicate with one another rather miraculous. On the naturalist-biological response to postmodern relativism, see Dennett (2009).

14. Green (2013) provides further examples of this in her discussion of the use of multiple models in systems biology.

15. In Van Fraassen (1980), and still more recently (2008), the empiricist position he defends maintains that we need not believe that good or successful scientific theories are true, only that they are empirically adequate in their accounts of observable phenomena.

16. In a similar vein, Eva Johach (2008, 11–12) has noted the hybrid nature of many scientific metaphors, so that their use is not simply a matter

of transferring ideas from one semantically or conceptually pure domain to another. The two domains brought into relation through a metaphor are often already hybrids resulting from previous cross-fertilizations. The epigraphical quotation from Pickens with which this book begins seems to make a similar point.

17. E.g., Toulmin (1960); Giere, Bickle, and Mauldin (2005); Giere (2006); Van Fraassen (2008); and Winther, forthcoming.

18. It should be noted, though, that the question of natural kinds—whether the world comes naturally parcelled up in specific packages—is distinct from the question of a natural (or objective) language, i.e., do those entities have their own proper names aside from the artificial ones we choose to give them? The second question is surely absurd, while not so the first. Hacking (1999, 96–99) describes Kuhn as having strongly nominalist views.

19. See Reynolds (2000) for discussion and response to common misconceptions of the Peircean account of truth.

20. According to Ledford (2015), some labs report unintended or "offsite" mutation rates as high as 60 percent using the CRISPR/Cas9 technique, depending on the source of the organism and the cell-line.

Bibliography

Abercrombie, M., and E. J. Ambrose. 1958. "Interference Microscope Studies of Cell Contacts in Tissue Culture." *Experimental Cell Research* 15: 322–45.

Abercrombie, M., and J. E. M. Heaysman. 1953. "Observations on the Social Behavior of Cells in Tissue Culture: I. Speed of Movement of Chick Heart Fibroblasts in Relation to Their Mutual Contacts." *Experimental Cell Research* 5(1): 111–31.

———. 1954a. "Observations on the Social Behavior of Cells in Tissue Culture II. 'Monolayering' of Fibroblasts." *Experimental Cell Research* 6: 293–306.

———. 1954b. "Invasiveness of Sarcoma Cells." *Nature* 174: 697–98.

Abercrombie, M., J. E. M. Heaysman, and H. M. Karthauser. 1957. "Social Behaviour of Cells in Tissue Culture III. Mutual Influence of Sarcoma Cells and Fibroblasts." *Experimental Cell Research* 13: 276–91.

Ackerknecht, Erwin H. 1953. *Rudolf Virchow: Doctor, Statesman, Anthropologist*. Madison: University of Wisconsin Press.

Aguilar, A. 1999. "The Key Action Cell Factory, an Initiative of the European Union." *International Microbiology* 2: 121–24.

Alberts, Bruce. 1998. "The Cell as a Collection of Protein Machines: Preparing the Next Generation of Molecular Biologists." *Cell* 92(3): 291–94.

Alberts, B., A. Johnson, J. Lewis, M. Raff, K. Roberts, and P. Walter. 2008. *Molecular Biology of the Cell*. 5th ed. New York: Garland.

Albrecht-Buehler, Guenter. 1990. "In Defense of 'Nonmolecular' Cell Biology." *International Review of Cytology* 120: 191–241.

Allen, Garland. 1978. *Life Science in the Twentieth Century*. Rev. ed. Cambridge: Cambridge University Press.

———. 2005. "Mechanism, Vitalism and Organicism in Late-Nineteenth and Twentieth-Century Biology: The Importance of Historical Context." *Studies in History and Philosophy of Biological and Biomedical Sciences* 36: 261–83.

———. 2007. "A Century of Evo-Devo: The Dialectics of Analysis and Synthesis in Twentieth-Century Life Science." In *From Embryology to Evo-Devo: A History of Developmental Biology*, edited by Manfred D. Laubichler and Jane Maienschein, 123–67. Cambridge, MA: MIT Press.

Ameisen, J. C. 2002. "On the Origin, Evolution, and Nature of Programmed Cell Death: A Timeline of Four Billion Years." *Cell Death and Differentiation* 9: 367–93.

Apter, Michael J. 1966. *Cybernetics and Development*. Oxford: Pergamon.

Apter, Michael J., and Lewis Wolpert. 1965. "Cybernetics and Development 1: Information theory." *Journal of Theoretical Biology* 8(2): 244–57.

Armon, Rony. 2012. "Between Biochemists and Embryologists—the Biochemical Study of Embryonic Induction in the 1930s." *Journal of the History of Biology* 45(1): 65–108.

Aschoff, Ludwig, Ernst Küster, and W. J. Schmidt. 1938. *Hundert Jahre Zellforschung*. Berlin: Gebrüder Borntraeger.

Baetu, Tudor. 2012a. "Emergence, Therefore Antireductionism? A Critique of Emergent Antireductionism." *Biology and Philosophy* 27(3): 433–48.

———. 2012b. "Genomic Programs as Mechanism Schemas: A Non-Reductionist Interpretation." *The British Journal for the Philosophy of Science* 63(3): 649–71.

———. 2014. "Models and the Mosaic of Scientific Knowledge: The Case of Immunology." *Studies in History and Philosophy of Biological and Biomedical Sciences* 45: 49–56.

Bailer-Jones, Daniela. 2002. "Models, Metaphors and Analogies." In *The Blackwell Guide to the Philosophy of Science*, edited by Peter Machamer and Michael Silberstein, 108–27. Malden, MA: Blackwell.

———. 2009. *Scientific Models in Philosophy of Science*. Pittsburgh: University of Pittsburgh Press.

Baker, John R. 1948–52. "The Cell-Theory: A Restatement, History, and Critique." *Quarterly Journal of Microscopical Science* Part I, vol. 89, 1948, Third Series, no. 1: 103–25; Part II, vol. 90, March 1949, part 1: 87–108; Part III, vol. 93, June 1952, part 2: 157–90.

Baldwin, Ernest. 1948. *An Introduction to Comparative Biochemistry*. 3rd ed. London: Cambridge University Press.

Ball, Philip. 2011. "A Metaphor Too Far." *Nature News*, 23 February. Accessed April 30, 2013. doi: 10.1038/news.2011.115.

Baluška, F., D. Volkmann, and P. W. Barlow. 2004. "Eukaryotic Cells and Their *Cell Bodies*: Cell Theory Revised." *Annals of Botany* 94: 9–32.

Barker, Gillian, and Philip Kitcher. 2014. *Philosophy of Science: A New Introduction*. New York: Oxford University Press.

Bartha, Paul. 2013. "Analogy and Analogical Reasoning." In *The Stanford Encyclopedia of Philosophy*. Fall ed. Edited by Edward N. Zalta. Accessed August 13, 2014. http://plato.stanford.edu/archives/fall2013/entries/reasoning-analogy/.

Bashor, Caleb J., Andrew A. Horwitz, Sergio G. Peisajovich, and Wendell A. Lim. 2010. "Rewiring Cells: Synthetic Biology as a Tool to Interrogate the Organizational Principles of Living Systems." *Annual Review of Biophysics* 39: 515–37.

Bassler, Bonnie, and Richard Losick. 2006. "Bacterially Speaking." *Cell* 125: 237–46.

Bayliss, Leonard E. 1959. *Principles of General Physiology*. Vol. 2. 5th ed. London: Longmans.

Bechtel, William. 2006. *Discovering Cell Mechanisms: The Creation of Modern Cell Biology*. Cambridge: Cambridge University Press.

———. 2010. "The Cell: Locus or Object of Inquiry?" *Studies in History and Philosophy of Biological and Biomedical Sciences* 41(3): 172–82.

Bechtel, William, and Adele Abrahamsen. 2005. "Explanation: A Mechanist Alternative." *Studies in History and Philosophy of Biological and Biomedical Sciences* 36: 421–41.

Beldecos, Athena, S. Bailey, S. Gilbert, K. Hicks, L. Kenschaft, N. Niemczyk, R. Rosenberg, S. Schaertel, and A. Wedel. 1988. "The Importance of Feminist Critique for Contemporary Cell Biology." *Hypatia* 3(1): 61–76.

Berg, Maxine. 1985. *The Age of Manufactures: 1700–1820*. Totowa, NJ: Barnes & Noble.

Bernard, Claude. [1865] 1961. *An Introduction to the Study of Experimental Medicine*. New York: Collier.

———. 1885. *Leçons sur les Phénomènes de la Vie communs aux Animaux et aux Végétaux*. Vol. 1. Paris: J-B. Baillière et Fils.

Bhalla, Upinder S. 2003. "Understanding Complex Signaling Networks through Models and Metaphors." *Progress in Biophysics & Molecular Biology* 81: 45–65.

Bhalla, Upinder S., and Ravi Iyengar. 1999. "Emergent Properties of Networks of Biological Signaling Pathways." *Science* 283: 381–87.

Bird, Adrian. 2002. "DNA Methylation Patterns and Epigenetic Memory." *Genes & Development* 16: 6–21.

Black, Max. 1962a. "Metaphor." In *Models and Metaphors: Studies in Language and Philosophy*, 25–47. Ithaca, NY: Cornell University Press.

———. 1962b. "Models and Archetypes." In *Models and Metaphors: Studies in Language and Philosophy*, 219–43. Ithaca, NY: Cornell University Press.

———. 1979. "More About Metaphor." In *Metaphor and Thought*, edited by Andrew Ortony, 19–41. Cambridge: Cambridge University Press.

Blasimme, Alessandro, Paolo Maugeri, and Pierre-Luc Germain. 2013. "What Mechanisms Can't Do: Explanatory Frameworks and the Function of the p53 Gene in Molecular Oncology." *Studies in History and Philosophy of Biological and Biomedical Sciences* 44(3): 374–84.

Blumenberg, Hans. 1981. *Die Lesbarkeit der Welt*. Frankfurt am Main: Suhrkamp.

———. 2010. *Paradigms for a Metaphorology*. Translated by Robert Savage. Ithaca: Cornell University Press.

Bolouri, Hamid, and Eric H. Davidson. 2010. "The Gene Regulatory Network Basis of the 'Community Effect,' and the Analysis of a Sea Urchin Embryo Example." *Developmental Biology* 340(2): 170–78.

Boniolo, Giovanni. 2013. "On Molecular Mechanisms and Contexts of Physical Explanation." *Biological Theory* 7: 256–65.

Bonner, James. 1965. *The Molecular Biology of Development*. New York: Oxford University Press.

Börlin, Christoph, Verena Lang, Anne Hamacher-Brady, and Nathan Brady. 2014. "Agent-Based Modeling of Autophagy Reveals Emergent Regulatory Behavior of Spatio-Temporal Autophagy Dynamics." *Cell Communication & Signaling* 12(56). Accessed November 4, 2014. doi. 10.1186/ s12964-014-0056-8.

Boudry, Maarten, and Massimo Pigliucci. 2013. "The Mismeasure of Machine: Synthetic Biology and the Trouble with Engineering Metaphors." *Studies in History and Philosophy of Biological and Biomedical Sciences* 44(4, B): 660–68.

Bourne, Gilbert C. 1896. "A Criticism of the Cell-Theory; Being an Answer to Mr. Sedgwick's Article on the Inadequacy of the Cellular Theory of Development." *Quarterly Journal of Microscope Science*, 38: 137–74.

Bowdle, Brian F., and Dedre Gentner. 2005. "The Career of Metaphor." *Psychological Review* 112(1): 193–216.

Boyd, Richard. 1993. "Metaphor and Theory Change: What is 'Metaphor' a Metaphor for?" In *Metaphor and Thought*. 2nd ed. Edited by Andrew Ortony, 481–532. Cambridge: Cambridge University Press.

Bradie, Michael. 1998. "Explanation as Metaphorical Redescription." *Metaphor and Symbol* 13(2): 125–39.

———. 1999. "Science and Metaphor." *Biology and Philosophy* 14: 159–66.

Braithwaite, R.B. [1953] 1960. *Scientific Explanation: A Study of the Function of Theory, Probability and Law in Science*. New York: Harper & Brothers.

Brandt, Christina. 2005. "Genetic Code, Text, and Scripture: Metaphors and Narration in German Molecular Biology." *Science in Context* 18(4): 629–48.

———. 2012. "Stem Cells, Reversibility and Reprogramming: Historical Perspectives." In *Differing Routes to Stem Cell Research: Germany and Italy*, edited by Renato Mazzolini and Hans-Jörg Rheinberger, 55–91. Berlin: Duncker & Humblot, Bologna: Societa editrice il Mulino.

Bray, Dennis. 1990. "Intracellular Signalling as a Parallel Distributed Process." *Journal of Theoretical Biology* 143: 215–31.

———. 2009. *Wetware: A Computer in Every Living Cell*. New Haven: Yale University Press.

Brenner, Sydney. 2012. "Life's Code Script." *Nature* 482(7386): 461.

Brown, James Robert. 2001. *Who Rules in Science: An Opinionated Guide to the Wars.* Cambridge, MA.: Harvard University Press.

Brown, Theodore L. 2003. *Making Truth: Metaphor in Science.* Urbana and Chicago: University of Illinois Press.

Buffon, Georges Louis Leclerc, Comte de. 1749. *Histoire Naturelle, Générale, et Particulière.* Vol. 2, *Histoire Générale des Animaux.* Paris: Imprimerie Royale.

Burke, Kenneth. 1965. *Permanence and Change.* Indianapolis: Bobbs-Merrill.

Cabeen, Matthew T., and Christine Jacobs-Wagner. 2010. "A Metabolic Assembly Line in Bacteria." *Nature Cell Biology* 12: 731–33.

Callebaut, Werner, ed. 1993. *Taking the Naturalistic Turn, or How Real Philosophy of Science Is Done.* Chicago: University of Chicago Press.

Camp, Elisabeth. 2006. "Metaphor and That Certain 'Je Ne Sais Quoi.'" *Philosophical Studies* 14: 159–66.

Canguilhem, Georges. 1963. "The Role of Analogies and Models in Biological Discovery." In *Scientific Change: Historical Studies in the Intellectual, Social and Technical Conditions for Scientific Discovery and Technical Invention, From Antiquity to the Present,* edited by A.C. Crombie, 507–20. London: Heinemann Educational Books Ltd.

———. 2008. "Cell Theory." In *Knowledge of Life,* edited by P. Marrati and T. Meyers. Translated by Stefanos Geroulanos and Daniela Ginsburg. New York: Fordham University Press.

Carpenter, William Benjamin. 1865. *A Manual of Physiology, Including Physiological Anatomy for the Use of the Medical Student,* 4th ed. London: John Churchill.

Carrel, A. 1931. "The New Cytology." *Science* 73(1890): 297–303.

———. 1937. *Man, the Unknown.* London: Hamish Hamilton.

Cartwright, Nancy. 1983. *How the Laws of Physics Lie.* New York: Oxford University Press.

———. 1999. *The Dappled World: A Study of the Boundaries of Science.* Cambridge: Cambridge University Press.

Cathcart, Edward Provan. 1928. "Dynamic Biochemistry." *Edinburgh Medical Journal* 35: 21–29.

Chakravarrty, Anjan. 2014. "Scientific Realism." *The Stanford Encyclopedia of Philosophy* (Spring 2014 ed.). Edited by Edward N. Zalta. http://plato.stanford.edu/archives/spr2014/entries/scientific-realism/ Accessed June 07, 2015.

Chambers, Robert. 1924. "The Physical Structure of Protoplasm as Determined by Micro-Dissection and Injection." In *General Cytology,* edited by Edmund V. Cowdry. 237–309, Chicago: University of Chicago Press.

Chandebois, Rosine. 1976. "Cell Sociology: A Way of Reconsidering the Current Concepts of Morphogenesis." *Acta Biotheoretica* 25(2–3): 71–102.

———. 1977. "Cell Sociology and the Problem of Position Effect: Pattern Formation, Origin and Role of Gradients." *Acta Biotheoretica* 26(4): 203–38.

———. 1980. "Cell Sociology and the Problem of Automation in the Development of Pluricellular Animals." *Acta Biotheoretica* 29(1): 1–35.

———. 1981. "The Problem of Automation in Animal Development:

Confrontation of the Concept of Cell Sociology with Biochemical Data." *Acta Biotheoretica* 30(1): 143–69.

Chandebois, Rosine, and Jacob Faber. 1983. *Automation in Animal Development: A New Theory Derived from the Concept of Cell Sociology.* New York: Karger.

———. 1987. "From DNA Transcription to Visible Structure: What the Development of Multicellular Animals Teaches Us." *Acta Biotheoretica* 36: 61–119.

Cheung, Tobias. 2010. "What Is an 'Organism'? On the Occurrence of a New Term and Its Conceptual Transformations 1680–1850." *History and Philosophy of the Life Sciences* 32(2–3): 155–94.

Chew, Matthew K., and Manfred D. Laubichler. 2003. "Natural Enemies— Metaphor or Misconception?" *Science* 301(5629): 52–53.

Churchill, Frederick B. 1989. "The Guts of the Matter. Infusoria from Ehrenberg to Bütschli: 1838–1876." *Journal of the History of Biology* 22(2): 189–213.

———. 2007. "Living with the Biogenetic Law: A Reappraisal." In *From Embryology to Evo-Devo: A History of Developmental Biology*, edited by Manfred D. Laubichler and Jane Maienschein, 37–81. Cambridge, MA: MIT Press.

Clarke, Peter G.H., and Stephanie Clarke. 2012. "Nineteenth Century Research on Cell Death." *Experimental Oncology* 34(3): 139–45.

Claude, Albert. 1948. "Studies on Cells: Morphology, Chemical Constitution and Distribution of Biochemical Functions." *Harvey Lectures* 43: 121–64.

Cleland, John. 1873. "On Cell Theories." *Quarterly Journal of Microscopical Science* 13(51): 255–66.

Clément, Pierre. 2007. "Introducing the Cell Concept with Both Animal and Plant Cells: A Historical and Didactic Approach." *Science and Education* 16: 423–40.

Coleman, William. 1977. *Biology in the Nineteenth Century: Problems of Form, Function, and Transformation.* Cambridge: Cambridge University Press.

Collins, James. 2012. "Synthetic Biology: Bits and Pieces Come to Life, Scientists Are Combining Biology and Engineering to Change the World." *Nature* 483 (1 March):S8–S10.

Conklin, Bruce R. 2007. "New Tools to Build Synthetic Hormonal Pathways." *Proceedings of the National Academy of Sciences* 104(12): 4777–78.

Contessa, Gabriele. 2011. "Scientific Models and Representation." In *The Continuum Companion to the Philosophy of Science*, edited by Steven French and Juha Saatsi, 120–37. London: Continuum.

Cramer, Friedrich. 1994. "Emil Fischer's Lock-and-Key Hypothesis After One Hundred Years—Towards a Supracellular Chemistry." In *The Lock-and-Key Principle, the State of the Art—100 Years On*, edited by Jean-Paul Behr. Chichester: John Wiley & Sons Ltd.

Danchin, Antoine. 2004. "The Bag or the Spindle: The Cell Factory at the Time of Systems' Biology." *Microbial Cell Factories* 3: 13.

———. 2012. "Scaling Up Synthetic Biology: Don't Forget the Chassis." *FEBS Letters* 586: 2129–37.

Daston, Lorraine, and Peter Galison. 2007. *Objectivity*. New York: Zone Books.

Davidson, Donald. 1984. "What Metaphors Mean." In *Inquiries into Truth and Interpretation*, 245–64. Oxford: Clarendon Press.

De Bary, Heinrich Anton. 1859. *Die Mycetozoen (Schleimpilze). Ein Beitrag zur Kenntnis die niedersten Organismen*. Leipzig: Wilhelm Engelmann.

———. 1879. Review of *Lehrbuch der Botanik für mittlere und höhere Lehranstalten* von K. Prantl, Leipzig, 1879. *Botanische Zeitung* (14), April 4, 1879: 221–23.

Dennett, Daniel. 2009. "Postmodernism and Truth." In *Philosophy: The Quest for Truth*. 7th ed., edited by Louis P. Pojman and Lewis Vaughn, 251–57, New York: Oxford University Press.

De Regt, Henk W. 2009. "Understanding in Scientific Explanation." In *Scientific Understanding: Philosophical Perspectives*, edited by Henk de Regt, Sabina Leonelli, and Kai Eigner, 21–42. Pittsburgh: University of Pittsburgh Press.

Dewey, John. [1920] 1954. *Reconstruction in Philosophy*. 5th ed. New York: Mentor.

———. [1929] 1960. *The Quest for Certainty: A Study of the Relation of Knowledge and Action*. New York: Capricorn Books.

D'Hombres, E. N. 2009. "'Un Organisme est une Société, et Réciproquement?' La Délimitation des Champs d'Extension des Sciences de la Vie et des Sciences Sociales chez Alfred Espinas (1877)." *Revue d'histoire des sciences* 62(2): 395–422.

———. 2012. "The 'Division of Labour': The Birth, Life and Death of a Concept." *Journal of the History of Biology* 45(1): 3–31.

Dobell, C. Clifford. 1911. "The Principles of Protistology." *Archiv für Protistenkunde* 23 (3): 269–310.

———. 1960. *Antony van Leeuwenhoek and His "Little Animals"; Being Some Account of the Father of Protozoology and Bacteriology and His Multifarious Discoveries in These Disciplines*. New York: Dover.

Doncaster, Leonard. 1920. *An Introduction to the Study of Cytology*. Cambridge: Cambridge University Press.

Dougherty, Edward R., and Michael L. Bittner. 2010. "Causality, Randomness, Intelligibility, and the Epistemology of the Cell." *Current Genomics* 11: 221–37.

Dröscher, Ariane. 2002. "Edmund B. Wilson's *The Cell* and Cell Theory between 1896 and 1925." *History and Philosophy of the Life Sciences* 24: 357–89.

———. 2008. "'Was ist eine Zelle?' Edmund B. Wilson's Diagramm als graphische Antwort." *Natur und Kultur* – Beiträge zur 15. Jahrestagung der DGGTB: 1–11.

———. 2012. "Where Does Stem Cell Research Stem from? A Terminological Analysis of the First Ninety Years." In *Differing Routes to Stem Cell Research: Germany and Italy*, edited by Renato Mazzolini and Hans-Jörg

Rheinberger, 19–54. Berlin: Duncker & Humblot, Bologna: Societa editrice il Mulino.

———. 2014a. "History of Cell Biology." In *Encyclopedia of Life Sciences (ELS)*. Chichester: John Wiley & Sons Ltd. Accessed June 17, 2014. doi: 10.1002/9780470015902.a0021786.pub2.

———. 2014b. "Images of Cell Trees, Cell Lines, and Cell Fates: The Legacy of Ernst Haeckel and August Weismann in Stem Cell Research." *History and Philosophy of the Life Sciences* 36(2): 157–86. Accessed July 25, 2014. doi: 10.1007/s40656-014-0028-8.

———. 2014c. "'*Lassen Sie mich die Pflanzenzelle als geschäftigen Spagiriker betrachten': Franz Ungers Beiträge zur Zellbiologie seiner Zeit.*" In *Einheit in der Diversität: Franz Ungers Naturforschung im internationalen Kontext*, edited by Marianne Klemun. Wien: Austrian Academy of Sciences.

Drysdale, John. 1874. *The Protoplasmic Theory of Life*. London: Baillière, Tindall, and Cox.

Duchesneau, Francois. 1987. *Genèse de la Théorie Cellulaire*. Montréal: Bellarmin.

Duhem, Pierre. [1906] 1991. *The Aim and Structure of Physical Theory*. Introduction by Jules Vuillemin. Princeton, NJ: Princeton University Press.

Dunker, A. Keith, J. David Lawson, Celeste J. Brown, Ryan M. Williams, Pedro Romero, Jeong S. Oh, Christopher J. Oldfield, Andrew M. Campen, Catherin M. Ratliff, Kerry W. Hipps, Juan Ausio, Mark S. Nissen, Raymond Reeves, ChulHee Kang, Charles R. Kissinger, Robert W. Bailey, Michael D. Griswold, Wah Chiu, Ethan C. Garner, and Zoran Obradovic. 2001. "Intrinsically Disordered Protein." *Journal of Molecular Graphics and Modelling* 19: 26–59.

Dunn, G., and G. Jones. 1998. "Michael Abercrombie: The Pioneer Ethologist of Cells." *Trends in Cell Biology* 8: 124–26.

Dutfield, Graham. 2012. "'The Genetic Code is 3.6 Billion Years Old: It's Time for a Re-Write': Questioning the Metaphors and Analogies of Synthetic Biology and Life Science Patenting." In *New Frontiers in the Philosophy of Intellectual Property*, edited by Annabelle Lever, 172–202. New York: Cambridge University Press.

Ede, D. A. 1972. "Cell Behavior and Embryonic Development." *International Journal of Neuroscience* 3: 165–74.

Edelman, Gerald M. 1988. *Topobiology: An Introduction to Molecular Embryology*. New York: Basic Books.

Ehrenberg, Christian. 1838. *Infusionthierchen als vollkomene Organismen*. Leipzig: Leopold Voss.

Elsberg, Louis. 1881. "The Cell-Doctrine and the Bioplasson-Doctrine." *Science* 2(76): 584–89.

Elwick, Jaime. 2003. "Herbert Spencer and the Disunity of the Social Organism." *History of Science* xli: 35–72.

———. 2007. *Styles of Reasoning in the British Life Sciences: Shared Assumptions, 1820–1858*. London: Pickering & Chatto Publishers.

———. 2014. "Containing Multitudes: Herbert Spencer, Organisms Social,

and Orders of Individuality." In *Herbert Spencer: Legacies*, edited by Mark
Francis and Michael W. Taylor, 89–110. Acumen Publishing (UK).

Ephrussi, Boris. 1953. *Nucleo-cytoplasmic Relations in Micro-organisms.*
Oxford: Clarendon Press.

Erickson, Brent, Rina Singh, and Paul Winters. 2011. "Synthetic Biology:
Regulating Industry Uses of New Biotechnologies." *Science* 333(2 September): 1254–56.

Fantuzzi, Giamila. 2017. "Cancer Is a Propagandist." *Studies in History and
Philosophy of Biological and Biomedical Sciences* 63: 28–31.

Farkas, I. J., T. Korcsmáros, I. A. Kovács, À. Mihalik, R. Palotai, K. Z. Szalay,
M. Szalay-Bekő, T. Vellai, S. Wang, and P. Csermely. 2011. "Network-Based
Tools for the Identification of Novel Drug Targets." *Science Signaling*
4(173): 1–5.

Feest, Uljana. 2010. "Concepts as Tools in the Experimental Generation of
Knowledge in Cognitive Psychology." *Spontaneous Generations: A Journal
for the History and Philosophy of Science* 4(1): 173–90.

Fehr, Carla. 2004. "Feminism and Science: Mechanism without Reductionism."
NWSA Journal 16(1): 136–56.

Feller, Stephan M. 2010. "The Dawn of a New Era in Cell Signalling Research." *Cell Communication and Signalling* 8: 7.

Fischer, A. 1923. "Contributions to the Biology of Tissue Cells. I. The Relation
of Cell Crowding to Tissue Growth In Vitro." *Journal of Experimental and
Biological Medicine* 38(6): 667–72.

Fischer, A., and A. B. Jensen. 1946. "Growth Limiting Factors of Tissue Cells
In Vitro." *Acta Physiologica Scandinavica* 12(1–2): 218–28.

Fisher, Michael J., Raymond C. Paton, and Koichiro Matsuno. 1999. "Intracellular Signalling Proteins as 'Smart' Agents in Parallel Distributed Processes." *BioSystems* 50: 159–71.

Flew, Antony. 1971. *Thinking About Thinking: Do I Sincerely Want to Be
Right?* London: Collins Fontana.

Forgacs, G. 1995. "On the Possible Role of Cytoskeletal Filamentous Networks in Intracellular Signaling: An Approach Based on Percolation."
Journal of Cell Science 108: 2131–43.

Foster, Michael. 1885. "Physiology." In *Encyclopedia Britannica*. Vol. XIX.
9th ed., 1878–1889, 8–62. New York: C. Scribner's Sons.

Friedman, Michael. 1974. "Explanation and Scientific Understanding." *Journal
of Philosophy* 71: 5–19.

Fruton, Joseph S. 1972. *Molecules and Life: Historical Essays on the Interplay
of Chemistry and Biology.* New York: Wiley & Sons.

———. 1999. *Proteins, Enzymes, Genes: The Interplay of Chemistry and Biology.* New Haven, CT: Yale University Press.

Garcia, Vivette, and Edna Suárez. 2010. "Switches and Batteries: Two Models
of Gene Regulation and a Note on the Historiography of 20th Century
Biology." In *The Hereditary Hourglass: Genetics and Epigenetics 1868–
2000.* Preprint 392, edited by Ana Barahona, Edna Suárez-Diaz, and Hans-
Jörg Rheinberger, 59–84. Berlin: Max Planck Institute for the History of
Science.

Gardel, Margaret L., Ian C. Schneider, Yvonne Aratyn-Schaus, and Clare M. Waterman. 2010. "Mechanical Integration of Actin and Adhesion Dynamics in Cell Migration." *Annual Review of Cell and Developmental Biology* 26: 315–33.

Gardner, Timothy S., Charles R. Cantor, and James J. Collins. 2000. "Construction of a Genetic Toggle Switch in *Escherichia coli*." *Nature* 403: 339–42.

Gass, Gillian L., and Jessica A. Bolker. 2007. "Modularity." In *Keywords and Concepts in Evolutionary Developmental Biology*, edited by B. K. Hall and W. M. Olson, 260–67. Cambridge, MA: Harvard University Press.

Gass, Gillian L., and B. K. Hall. 2007. "Collectivity in Context: Modularity, Cell Sociology, and the Neural Crest." *Biological Theory* 2(4): 349–59.

Geison, Gerald L. 1969. "The Protoplasmic Theory of Life and the Vitalist-Mechanist Debate." *Isis* 60: 273–92.

Gentner, Dedre. 1982. "Are Scientific Metaphors Analogies?" In *Metaphor: Problems and Perspectives*, edited by David S. Miall, 106–32. Sussex: Harvester Press, Ltd.

———. 1983. "Structure-Mapping: A Theoretical Framework for Analogy." *Cognitive Science* 7: 155–70.

Gentner, Dedre, and Michael Jeziorski. 1993. "The Shift from Metaphor to Analogy in Western Science." In *Metaphor and Thought*. 2nd ed., edited by Andrew Ortony, 447–80. Cambridge: Cambridge University Press.

Gibson, Toby J. 2009. "Cell Regulation: Determined to Signal Discrete Cooperation." *Trends in Biochemical Sciences* 34(10): 471–82.

Giere, Ronald N. 2006. *Scientific Perspectivism*. Chicago: University of Chicago Press.

Giere, Ronald N., John Bickle, and Robert F. Mauldin. 2005. *Understanding Scientific Reasoning*. 5th ed. Toronto: Thomson.

Gilbert, Scott F. 1992. "Cells in Search of Community: Critiques of Weismannism and Selectable Units in Ontogeny." *Biology and Philosophy* 7: 473–87.

———. 2003. *Developmental Biology*. 7th ed. Sunderland, MA: Sinauer.

Gilbert, Scott F., and S. Sarkar. 2000. "Embracing Complexity: Organicism for the 21st Century." *Developmental Dynamics* 219: 1–9.

Giuseppin, Marco L. F., Theo C. Verrips, and Natal A. W. van Riel. 1999. "The Cell Factory Needs a Model of a Factory." *Trends in Biotechnology* 17 (October): 383–84.

Glennan, Stuart. 1996. "Mechanisms and the Nature of Causation." *Erkenntnis* 44: 49–71.

Gliboff, Sander. 2008. *H. G. Bronn, Ernst Haeckel, and the Origins of German Darwinism: A Study in Translation and Transformation*. Cambridge, MA: MIT Press.

Godfrey-Smith, Peter. 2003. *Theory and Reality: An Introduction to the Philosophy of Science*. Chicago: University of Chicago Press.

———. 2009. *Darwinian Populations and Natural Selection*. Oxford: Oxford University Press.

Gomperts, Bastien D., Ijsbrand M. Kramer, and Peter E. R. Tatham. 2009. *Signal Transduction*. 2nd ed. Burlington, MA: Academic Press.

Goodman, Nelson. 1979. "Metaphor as Moonlighting." *Critical Inquiry* 6(1): 125–30.

Goschler, Constantin. 2002. *Rudolf Virchow: Mediziner—Anthropologe—Politiker*. Köln, Weimar, Wien: Böhlau Verlag.

Gould, Stephen J. 1995. "Four Metaphors in Three Generations." In *Dinosaur in a Haystack: Reflections on Natural History*, 442–57. New York: Harmony.

Grant, James. 2010. "The Dispensability of Metaphor." *British Journal of Aesthetics* 50(3): 255–72.

Gray, James. 1931. *A Textbook of Experimental Cytology*. Cambridge: Cambridge University Press.

Green, David E. 1937. "Reconstruction of the Chemical Events in Living Cells." In *Perspectives in Biochemistry: Thirty-One Essays Presented to Sir Frederick Gowland Hopkins by Past and Present Members of His Laboratory*, edited by Joseph Needham and David. E. Green, 175–86. London: Cambridge University Press.

Green, Sara. 2013. "When One Model Is Not Enough: Combining Epistemic Tools in Systems Biology." *Studies in History and Philosophy of Biological and Biomedical Sciences* 44(2): 170–80.

Griffiths, Paul E. 2001. "Genetic Information: A Metaphor in Search of a Theory." *Philosophy of Science* 68(3): 394–412.

Grimm, Stephen. 2010. "The Goal of Explanation." *Studies in History and Philosophy of Science* 41: 337–44.

Grote, Mathias. 2010. "Surfaces of Action: Cells and Membranes in Electrochemistry and the Life Sciences." *Studies in History and Philosophy of Biological and Biomedical Sciences* 41(3): 183–93.

Guilluy, Christophe, Rafel Garcia-Mata, and Keith Burridge. 2011. "Rho Protein Crosstalk: Another Social Network?" *Trends in Cell Biology* 21(12): 718–26.

Gupta, Shakti, Mano Ram Maurya, and Shankar Subramaniam. 2010. "Identification of Crosstalk Between Phosphoprotein Signaling Pathways in RAW 264.7 Macrophage Cells." *PloS Computational Biology* 6(1) e1000654. Accessed August 13, 2012. doi: 10.1371/journal.pcbi.1000654.

Gurdon, J. B. 1988. "A Community Effect in Animal Development." *Nature* 336(29 December): 772–74.

Hacking, Ian. 1983. *Representing and Intervening: Introductory Topics in the Philosophy of Natural Science*. Cambridge: Cambridge University Press.

———. 1999. *The Social Construction of What?* Cambridge MA: Harvard University Press.

Haeckel, Ernst. 1866. *Die Generelle Morphologie der Organismen. Allgemeine Grundzüge der Organischen Formen-Wissenschaft, mechanisch begründet durch die von Charles Darwins reformierte Deszendenz-Theorie.* 2 vols. Berlin: G. Reimer.

———. 1868. *Natürliche Schöpfungsgeschichte. Gemeinverständliche wissenschaftliche Vorträge über die Entwicklungslehre im Allgemeinen und diejenige von Darwin, Goethe und Lamarck im Besonderen, über die*

Anwendung derselben auf den Ursprung des Menschen und andere damit zussamenhängende Grundfragen der Naturwissenschaften. 2 vols. Berlin: G. Reimer.

———. 1869. *Zur Entwickelungsgeschichte der Siphonophoren, Beobachtungen über die Entwickelungsgeschichte der Genera Physophora, Crystallodes, Athorybi, und Reflexionen über die Entwickelungsgeschichte der Siphonophoren im Allgemeinen. Eine von der Utrechter Gesellschaft für Kunst und Wissenschaft gekrönte Preisschrift, mit vierzehn Tafeln*. Utrecht.

———. [1876] 1879. "Ueber die Wellenzeugung der Lebenstheilchen oder die Perigenesis der Plastidule." Reprinted in *Gesammelte populäre Vorträge aus dem Gebiete der Entwickelungslehre*. vol. 2. Bonn.

Haldane, John B. S. 1937. "The Biochemistry of the Individual." In *Perspectives in Biochemistry: Thirty-One Essays Presented to Sir Frederick Gowland Hopkins by Past and Present Members of His Laboratory*, edited by Joseph Needham and David. E. Green, 1–10. London: Cambridge University Press.

Hall, Brian K. 2000. "The Neural Crest as a Fourth Germ Layer and Vertebrates as Quadroblastic not Triploblastic." *Evolution & Development* 2: 1–3.

———. 2003. "Unlocking the Black Box between Genotype and Phenotype: Cell Condensations as Morphogenetic (modular) Units." *Biology and Philosophy* 18: 219–47.

Hall, Brian K., and Tsutomu Miyake. 1995. "Divide, Accumulate, Differentiate: Cell Condensation in Skeletal Development Revisited." *International Journal of Developmental Biology* 39: 881–93.

———. 2000. "All for One and One for All: Condensations and the Initiation of Skeletal Development." *BioEssays* 22(2): 138–47.

Hall, Thomas S. 1969. *Ideas of Life and Matter*. Vol. 2. Chicago: University of Chicago Press.

Hanahan, Douglas, and Robert A. Weinberg. 2000. "The Hallmarks of Cancer." *Cell* 100: 57–70.

———. 2011. "Hallmarks of Cancer: The Next Generation." *Cell* 144: 646–74.

Hänseler, Marianne. 2009. *Metaphern unter dem Mikroskop: Die epistemische Rolle von Metaphorik in den Wissenschaften und in Robert Kochs Bakteriologie*. Zurich: Chronos.

Hanson, Norwood Russell. 1958. *Patterns of Discovery: An Inquiry into the Conceptual Foundations of Science*. Cambridge: Cambridge University Press.

Haraway, Donna Jeanne. [1976] 2004. *Crystals, Fabrics, and Fields: Metaphors that Shape Embryos*. Berkeley, California: North Atlantic Books.

Haroske, G., V. Dimmer, D. Steindorf, U. Schilling, F. Theissig, and K. D. Kunze. 1996. "Cellular Sociology of Proliferating Tumour Cells in Invasive Ductal Breast Cancer." *Analytical and Quantitative Cytology and Histology* 18(3): 191–98.

Harré, Rom. 1972. *The Philosophies of Science: An Introductory Survey*. London: Oxford University Press.

Harris, Henry. 1999. *The Birth of the Cell*. New Haven: Yale University Press.

Hartwell, L., J. J. Hopfield, S. Leibler, and A. Murray. 1999. "From Molecular to Modular Cell Biology." *Nature* 402 (6761 Suppl.):C47–C52.

Hazen, Robert M., and James Trefil. 2009. *Science Matters: Achieving Scientific Literacy*. New ed. New York: Anchor Books.

Heidenhain, Martin. 1907. *Plasma und Zelle. Allgemeine Anatomie der lebendigen Masse*. Jena: Fischer.

Heitzmann, Carl. 1883. *Microscopical Morphology of the Animal Body in Health and Disease*. New York: J. H. Vail & Co.

Hempel, Carl G. 1965. *Aspects of Scientific Explanation and Other Essays in the Philosophy of Science*. New York: The Free Press.

Henderson, John. 2005. "Ernest Starling and 'Hormones': An Historical Commentary." *Journal of Endocrinology* 184: 5–10.

Heppner, G. H. 1993. "Cancer Cell Societies and Tumor Progression." *Stem Cells* 11: 199–203.

Hertwig, Oscar. 1895. *The Cell: Outlines of General Anatomy and Physiology*. Edited by Henry Johnstone Campbell M.D. Translated by M. Campbell. London: Sonnenschein & Co.

Hesse, Mary B. 1966. *Models and Analogies in Science*. Notre Dame: University of Notre Dame Press.

———. 1980. *Revolutions and Reconstructions in the Philosophy of Science*. Bloomington: Indiana University Press.

Hofmeister, Franz. 1901. *Die chemische Organisation der Zelle*. Braunschweig: Friedrich Wieweg und Sohn.

Holton, Gerald. 1995. "Metaphors in Science and Education." In *From a Metaphorical Point of View: A Multidisciplinary Approach to the Cognitive Content of Metaphor*, edited by Zdravko Radman, 259–88. Berlin: De Gruyter.

Holyoak, K. J., and Paul Thagard. 1995. *Mental Leaps: Analogy in Creative Thought*. Cambridge, MA: MIT Press.

Hooke, Robert. [1665] 2007. *Micrographia: Some Physiological Descriptions of Minute Bodies Made by Magnifying Glasses with Observations and Inquiries Thereupon*. Bibliobazaar.

Hope, Ronen, Efrat Ben-Mayor, Nehemya Friedman, Konstantin Voloshin, Dipul Biswas, Devorah Matas, Yaron Drori, Arthur Günzl, and Shulamit Michaeli. 2014. "Phosphorylation of the TATA-Binding Protein Activates the Spliced Leader Silencing Pathway in *Trypanosoma brucei*." *Science Signaling* 7 (2 September), 341. Accessed Sept. 3, 2014.

Hopkins, Francis G. 1913. "The Dynamic Side of Biochemistry." In *Report of the British Association for the Advancement of Science*, 652–68. London: John Murray

Hughes, Arthur. 1959. *A History of Cytology*. London: Abelard-Schuman.

Huxley, Thomas H. 1853. "The Cell Theory." *British and Foreign Medico-Chirurgical Review* 12: 285–314.

———. [1868] 1968. "On the Physical Basis of Life." *Collected Essays*, Vol. 1, 130–65. New York: Greenwood Press.

Hynes, Nancy E., Philip W. Ingham, Wendell A. Lim, Christopher J. Marshall, Joan Massagué, and Tony Pawson. 2013. "Signalling Change: Signal

Transduction Through the Decades." *Nature Reviews Molecular Cell Biology* 14(June): 393–98. Accessed June 30, 2014. doi: 10.1038/nrm3581.

Ishizaki, Y., L. Cheng, A.W. Mudge, and M. C. Raff. 1995. "Programmed Cell Death by Default in Embryonic Cells, Fibroblasts, and Cancer Cells." *Molecular Biology of the Cell* 6: 1443–58.

Jacob, François. 1979. "The Switch." In *Origins of Molecular Biology: A Tribute to Jacques Monod*, edited by André Lwoff and Agnes Ullmann, 95–108. New York: Academic Press.

Jacobs, Natasha X. 1989. "From Unit to Unity: Protozoology, Cell Theory, and the New Concept of Life." *Journal of the History of Biology*. 22(2): 215–42.

Jacyna, L. Stephen. 1984. "The Romantic Programme and the Reception of Cell Theory in Britain." *Journal of the History of Biology*, 17(1): 13–48.

———. 1990. "Romantic Thought and the Origins of Cell Theory." In *Romanticism and the Sciences*, edited by Andrew Cunningham and Nicholas Jardine, 161–69. Cambridge: Cambridge University Press.

Johach, Eva. 2008. *Krebszelle und Zellenstaat: zur medizinischen und politischen Metaphorik in Rudolf Virchows Zellularpathologie*. Berlin: Rombach Verlag KG.

Jordan, J. Dedrick, Emmanual M. Landau, and Ravi Iyengar. 2000. "Signaling Networks: The Origins of Cellular Multitasking." *Cell* 103: 193–200.

Jorgensen, Claus, and Rune Linding. 2010. "Simplistic Pathways or Complex Networks?" *Current Opinion in Genetics & Development* 20: 15–22.

Kalckar, H. M. 1965. "Galactose Metabolism and Cell 'Sociology.'" *Science* 150(3694): 305–13.

Kaltschmidt, Julia A., and Alfonso Martinez Arias. 2002. "A New Dawn for an Old Connection: Development Meets the Cell." *Trends in Cell Biology* 12(7): 316–20.

Kant, Immanuel. 1992 [1781]. *Critique of Pure Reason*. Translated by Norman Kemp Smith. Hong Kong: MacMillan.

Kaplan, Donald. R., and Wolfgang Hagemann. 1991. "The Relationship of Cell and Organism in Vascular Plants: Are Cells the Building Blocks of Plant Form?" *Bioscience*, 41(10): 693–703.

Karzbrun, E., A. M. Tayar, V. Noireaux, and E. H. Bar-Ziv. 2014. "Programmable On-Chip DNA Compartments as Artificial Cells." *Science* 345(6198): 829–32.

Karr, Jonathan R., Jayodita C. Sanghvi, Derek N. Macklin, Miriam V. Gutschow, Jared M. Jacobs, Benjamin Bolival Jr., Nacrya Assad-Garcia, John I. Glass, and Markus W. Covert. 2012. "A Whole-Cell Computational Model Predicts Phenotype from Genotype." *Cell* 150(2): 389–401.

Katz, Bernhard. 1950. "Depolarization of Sensory Terminals and the Initiation of Impulses in the Muscle Spindle." *Journal of Physiology* 111: 261–82.

Kay, Lily. 2000. *Who Wrote the Book of Life? A History of the Genetic Code*. Stanford: Stanford University Press.

Keith, Sir Arthur. 1924. "Does Man's Body Represent a Commonwealth?" In *The R. P. A. Annual for the Year 1924*, edited by Charles A. Watts, 3–12. London: Watts & Co.

Keller, Evelyn Fox. 1995. *Refiguring Life: Metaphors of Twentieth-Century Biology*. New York: Columbia University Press.

———. 2000. *The Century of the Gene*. Cambridge, MA: Harvard University Press.

———. 2002. *Making Sense of Life: Explaining Biological Development with Models, Metaphors, and Machines*. Cambridge, MA: Harvard University Press.

———. 2009a. "Knowing as Making, Making as Knowing: The Many Lives of Synthetic Biology." *Biological Theory* 4(4): 333–39.

———. 2009b. "What Does Synthetic Biology Have to Do with Biology?" *Biosocieties* 4(2–3): 291–302.

Kennedy, E. P. 1996. "Herman Moritz Kalckar." In *Biographical Memoirs*. Vol. 69, 149–66. Washington, DC: The National Academies Press.

Kerr, John F. R., Andrew H. Wyllie, and A. R. Currie. 1972. "Apoptosis: A Basic Biological Phenomenon with Wide-Ranging Implications in Tissue Kinetics." *British Journal of Cancer* 26: 239–57.

Khalifa, Kareem. 2012. "Inaugurating Understanding or Repackaging Explanation?" *Philosophy of Science* 79: 15–37.

Kiss, R., I. Camby, I. Salmon, P. van Ham, J. Brotchi, and J-L Pasteels. 1995. "Relationship between DNA Ploidy Level and Tumor Sociology Behavior in 12 Nervous Cell Lines." *Cytometry* 20: 118–126.

Kitcher, Philip. 1989. "Explanatory Unification and the Causal Structure of the World." In *Scientific Explanation*, edited by Philip Kitcher and Wesley Salmon, 410–505. Minneapolis: University of Minnesota Press.

Kittay, Eva Feder. 1987. *Metaphor: Its Cognitive Force and Linguistic Structure*. Oxford: Clarendon Press.

Kohler, Robert E. 1982. *From Medical Chemistry to Biochemistry: The Making of a Biomedical Discipline*. Cambridge: Cambridge University Press.

Korcsmáros, Tamás, Illés J. Farkas, Máté S. Szalay, Petra Rovó, David Fazekas, Zoltán Spiró, Csaba Böde, Katalin Lenti, Tibor Vellai, and Péter Csermely. 2010. "Uniformly Curated Signaling Pathways Reveal Tissue-Specific Cross-Talks and Support Drug Target Discovery." *Bioinformatics* 26: 2042–50.

Kordon, Claude. 1993. *The Language of the Cell*. Translated by William J. Gladstone. New York: McGraw-Hill Inc.

Koschorke, Albrecht, Susanne Lüdemann, Thomas Frank, and Ethel Matala de Mazza. 2007. *Der fiktive Staat: Konstruktionen des politischen Körpers in der Geschichte Europas*. Frankfurt am Main: Fischer Taschenbuch Verlag.

Kuhn, Thomas S. 1993. "Metaphor in Science." In *Metaphor and Thought*. 2nd ed., edited by A. Ortony, 533-42. Cambridge: Cambridge University Press.

———. 2012. *The Structure of Scientific Revolutions. 50th Anniversary Edition with an Introductory Essay by Ian Hacking*. Chicago: University of Chicago Press.

Kyne, Ciara, and Peter B. Crowley. 2016. "Grasping the Nature of the Cell Interior: From Physiological Chemistry to Chemical Biology." *Federation of European Biochemical Societies (FEBS) Journal* 283(16): 3016–28.

Lakoff, George. 1987. "The Death of Dead Metaphor." *Metaphor and Symbolic Activity* 2(2): 143–47.

Lakoff, George, and Mark Johnson. 2003 [1980]. *Metaphors We Live By.* Second Impression with a New Afterword. Chicago: University of Chicago Press.

Landecker, Hannah. 2007. *Culturing Life: How Cells Became Technologies.* Cambridge, MA: Harvard University Press.

———. 2011. "Creeping, Drinking, Dying: The Cinematic Portal and the Microscopic World of the Twentieth-Century Cell." *Science in Context* 24(3): 381–416.

———. 2016. "The Social as Signal in the Body of Chromatin." *The Sociological Review Monographs* 64(1): 79–99.

Lanphier, Edward, Fyodor Urnov, Sarah Ehlen Haecker, Michael Werner, and Joanna Smolenski. 2015. "Don't Edit the Human Germline." *Nature* 519: 410–11.

Lazebnik, Yuri. 2002. "Can a Biologist Fix a Radio? – Or, What I Learned While Studying Apoptosis." *Cancer Cell* 2(3): 179–82.

Leclercq, J., and J. E. Dumont. 1983. "Boolean Analysis of Cell Regulation Networks." *Journal of Theoretical Biology* 104: 507–34.

Ledford, Heidi. 2015. "CRISPR, the Disruptor." *Nature, News Feature*, 03 June 2015. http://www.nature.com/news/crispr-the-disruptor-1.17673. Accessed June 3, 2015.

Lehoux, Darren. 2006. "Laws of Nature and Natural Laws." *Studies in History and Philosophy of Science* 37: 527–49.

Lenoir, Timothy. 1989. *The Strategy of Life: Teleology and Mechanics in Nineteenth-Century German Biology.* Chicago: University of Chicago Press.

Levine, Michael, and Eric Davidson. 2005. "Gene Regulatory Networks for Development." *Proceedings of the National Academy of Sciences of the United States of America (PNAS)* 102(14): 4936–42.

Lewitzky, M., P. C. Simister, and S. M. Feller. 2012. "Beyond 'Furballs' and 'Dumpling Soups' —Towards a Molecular Architecture of Signaling Complexes and Networks." *FEBS Letters* 586: 2740–50.

Lewontin, Richard. 1991. *Biology as Ideology: The Doctrine of DNA.* Toronto: Anansi.

———. 2000. *The Triple Helix: Gene, Organism, and Environment.* Cambridge: Harvard University Press.

———. 2001. "In the Beginning Was the Word." Review of *Who Wrote the Book of Life? A History of the Genetic Code*, by Lily Kay. *Science* 291(5507): 1263–64.

Leydig, Franz. 1857. *Lehrbuch der Histologie des Menschen und Thiere.* Frankfurt am Main: Meidinger Sohn und Co.

Liddel, H. G., and R. Scott. 1989. *An Intermediate Greek-English Lexicon.* Oxford: Clarendon Press.

Limoges, Camille. 1994. "Milne-Edwards, Darwin, Durkheim and the Division of Labour: A Case Study in the Conceptual Exchanges Between the Social and the Natural Sciences." In *The Natural Sciences and the Social Sciences,* edited by I. B. Cohen, 317–43. Boston Studies in the Philosophy of Science, Dordrecht: Kluwer.

Lin, Chi-Hung, and Paul Forscher. 1995. "Growth Cone Advance Is Inversely Proportional to Retrograde F-Actin Flow." *Neuron* 14: 763–71.

Liu, Jing, and Anning Lin. 2007. "Wiring the Cell Signaling Circuitry by the NF-κB and JNK1 Crosstalk and Its Applications in Human Diseases." *Oncogene* 26: 3267–78.

Lockshin, Richard A., and Carroll M. Williams, 1964. "Programmed Cell Death—II. Endocrine Potentiation of the Breakdown of Intersegmental Muscles of Silkmoths." *Journal of Insect Physiology* 10: 643–49.

Loeb, Jacques. 1912. *The Mechanistic Conception of Life: Biological Essays.* Chicago: University of Chicago Press.

———. 1915. "Mechanistic Science and Metaphysical Romance." *Yale Review* 4: 766–85.

———. 1916. *The Organism as a Whole: From a Physicochemical Viewpoint.* New York: G. P. Putnam's Sons.

Loettgers, Andrea. 2013. "Communication: Metaphors Advance Scientific Research." *Nature* 502: 303.

Loewenstein, Werner R. 1960. "Biological Transducers." *Scientific American* 203(2): 98–108.

———. 1964. "Permeability of Membrane Junctions." *Annals of the New York Academy of Sciences* 137(2): 441–72.

———. 1965. "Facets of a Transducer Process." *Cold Spring Harbor Symposia on Quantitative Biology* 30: 29–43.

Loewenstein, W. R., and Y. Kanno. 1966. "Intercellular Communication and the Control of Tissue Growth: Lack of Communication Between Cancer Cells." *Nature* 209: 1248–49.

Longino, Helen. 1990. *Science as Social Knowledge: Values and Objectivity in Scientific Inquiry.* Princeton, NJ: Princeton University Press.

Ma, Huan, Rachel D. Groth, Samuel M. Cohen, John F. Emery, Boxing Li, Esthelle Hoedt, Guoan Zhang, Thomas A. Nuebert, and Richard W. Tsien. 2014. "γCaMKII Shuttles Ca^{2+}/CaM to the Nucleus to Trigger CREB Phosphorylation and Gene Expression." *Cell* 159(2): 281–94.

Maasen, Sabine, Everett Mendelsohn, and Peter Weingart. *Biology as Society, Society as Biology: Metaphors.* Boston: Dordrecht.

MacDonald, A.M., ed. 1966. *Chambers's Etymological English Dictionary.* New Edition with Supplement. London: W. & R. Chambers, Ltd.

Machamer, Peter, Lindley Darden, and Carl Craver. 2000. "Thinking About Mechanisms." *Philosophy of Science* 67(1): 1–25.

MacNichol, E. F. Jr. 1956. "Visual Receptors as Biological Transducers." In *Molecular Structure and Functional Activity of Nerve Cells: A Symposium Organized by the American Institute of Biological Sciences and Sponsored by the Office of Naval Research,* held in Washington, DC, June 3–4, 1955. Publ. no. 1, edited by Robert G. Grenell and L. J. Mullins, 34–52. Washington, DC: American Institute of Biological Sciences.

Maehle, Andreas-Holger. 2009. "A Binding Question: The Evolution of the Receptor Concept." *Endeavour* 33(4): 134–39.

————. 2011. "Ambiguous Cells: The Emergence of the Stem Cell Concept in the Nineteenth and Twentieth Centuries." *Notes & Recordings of the Royal Society of London* 65(4): 359–78.

Maienschein, Jane. 1991a. "Cytology in 1924: Expansion and Collaboration." In *The Expansion of American Biology*, edited by Keith R. Benson, Jane Maienschein, and Ronald Rainger, 23–51. New Brunswick, NJ: Rutgers University Press.

————. 1991b "From Presentation to Representation in E. B. Wilson's The Cell." *Biology and Philosophy* 6: 227–54.

Marcelpoil, R., and Y. Usson. 1992. "Methods for the Study of Cellular Sociology: Voronoï Diagrams and Parametrization of the Spatial Relations." *Journal of Theoretical Biology* 154: 359–69.

Marcelpoil, R., E. Beaurepaire, and S. Pesty. 1994. "La Sociologie Cellulaire: Modéliser et Simuler une 'Société' Cellulaire Pour Étudier le Vivant." *Intellectica* 2(19): 53–72.

Marks, Friedrich, Ursula Klingmüller, and Karin Müller-Decker. 2009. *Cellular Signaling Processing: An Introduction to the Molecular Mechanisms of Signal Transduction*. New York: Garland Science.

Martin, Emily. 1990. "Toward an Anthropology of Immunology: The Body as Nation State." *Medical Anthropology Quarterly* 4(4): 410–26.

————. 1991. "The Egg and the Sperm: How Science Has Constructed a Romance Based on Stereotypical Male-Female Roles." *Signs: Journal of Women in Culture and Society* 16(3): 485–501.

Matlin, Karl. 2002. "The Strange Case of the Signal Recognition Particle." *Nature Review Molecular Cell Biology* 3: 538–42.

Matlin, Karl, Jane Maienschein, and Manfred Laubichler, eds. 2018. *Visions of Cell Biology: Reflections Inspired by Cowdry's "General Cytology."* Chicago: University of Chicago Press.

Maxwell, Grover. [1962] 2013. "The Ontological Status of Theoretical Entities." Reprinted in *Philosophy of Science: The Central Issues. 2nd ed.*, edited by Martin Curd, J. A. Cover, and Christopher Pincock, 1049–59. New York: W. W. Norton & Co. Originally published in Herbert Feigel and Grover Maxwell, eds., *Scientific Explanation, Space, and Time*. Vol. 3, *Minnesota Studies in the Philosophy of Science*. Minneapolis: University of Minnesota Press, 1962.

Mayer, Bruce J., Michael L. Blinov, and Leslie M. Loew. 2009. "Molecular Machines or Pleiomorphic Ensembles: Signaling Complexes Revisited." *Journal of Biology* 8(9): 81. Accessed August 23, 2010. http://jbiol.com/content/8/9/81.

Mazzarello, Paolo. 1999. "A Unifying Concept: The History of Cell Theory." *Nature Cell Biology* 1 (May 1999): E13–15.

Mazzolini, Renato G. 1988. *Politisch-biologische Analogien im Frühwerk Rudolf Virchows*. Marburg: Basislken-Presse.

McMullin, Ernan. 1984. "A Case for Scientific Realism." In *Scientific Realism*, edited by Jarrett Leplin, 8–40. Berkeley: University of California Press.

Medawar, P. B. 1980. "Michael Abercrombie. 14 August 1912–28 May 1979." *Biographical Memoirs of the Fellows of the Royal Society*, 26: 1–15.

Meinesz, Alexandre. 2008. *How Life Began: Evolution's Three Geneses*. Chicago: University of Chicago Press.

Mendelsohn, J. Andrew. 2003. "Lives of the Cell." *Journal of the History of Biology* 36: 1–37.

Merz, J. T. 1965. *A History of European Thought in the Nineteenth Century*. Vol. 2. New York: Dover.

Miller, Arthur I. 1996. *Insights of Genius: Imagery and Creativity in Science and Art*. New York: Springer-Verlag.

Milne-Edwards, Henri. 1851. *Introduction à la Zoologie Générale, ou, Considérations sur les tendances de la nature dans la constitution du regne animal*. Paris: Victor Masson.

Mitchell, Peter. 1979. "Keilin's Respiratory Chain Concept and its Chemiosmotic Consequences." *Science* 206(4423): 1148–59.

———. 1991. "Foundations of Vectorial Metabolism and Osmochemistry." *Bioscience Reports* 11(6): 297–346.

Moberg, Carol L. 2012. *Entering an Unseen World: A Founding Laboratory and Origins of Modern Cell Biology 1910–1974*. New York: Rockefeller University Press.

Montgomery, Edmund. 1881. "Are We 'Cell-Aggregates'?" *Mind* 7(25): 100–107.

Morange, Michel. 2000. *A History of Molecular Biology*. Translated by Matthew Cobb. Cambridge, MA: Harvard University Press.

Moskowitz, Steven, and William C. Aird. 2007. "Shall I Compare the Endothelium to a Summer's Day? The Role of Metaphor in Communicating Science." In *Endothelial Biomedicine*, edited by William C. Aird, 199–210. New York: Cambridge University Press.

Moss, Lenny. 2012. "Is the Philosophy of Mechanism Philosophy Enough?" *Studies in History and Philosophy of Biological and Biomedical Sciences* 43: 164–172.

Müller-Wille, Staffan. 2010. "Cell Theory, Specificity, and Reproduction, 1837–70." *Studies in History and Philosophy of Biological and Biomedical Sciences C* 41(3): 225–31.

Nagel, Thomas. 1986. *The View from Nowhere*. New York: Oxford University Press.

Nakayama, N., Y. Kaziro, K. Arai, and K. Matsumoto. 1988. "Role of STE Genes in the Mating Factor Signaling Pathway Mediated by GPA1 in *Saccharomyces cerevisiae*." *Molecular and Cell Biology*, Sept. 8(9): 3777–83.

Nanney, D. 1989. "Metaphor and Mechanism: 'Epigenetic Control Systems' Reconsidered." Paper presented at Symposium on "The Epigenetics of Cell Transformation and Tumor Development," American Association for Cancer Research, Eightieth Annual Meeting, San Francisco, CA, May 26.

Nathan, Marco J. 2014. "Molecular Ecosystems." *Biology and Philosophy* 29: 101–22.

Nedelcu, A. M., W. W. Driscoll, P. M. Durand, M. D. Herron, and A. Rashidi. 2010. "On the Paradigm of Altruistic Suicide in the Unicellular World." *Evolution* 65(1): 3–20.

Nelkin, Dorothy. 1994. "Promotional Metaphors and Their Popular Appeal." *Public Understanding of Science* 3(1): 25–31.

Nicholson, Daniel J. 2010. "Biological Atomism and Cell Theory." *Studies in History and Philosophy of Biological and Biomedical Sciences* 41(3): 202–11.

———. 2013. "Organisms ≠ Machines." *Studies in History and Philosophy of Biological and Biomedical Sciences* 44 (4, B): 669–78.

———. 2014a. "The Machine Conception of the Organism: A Critical Analysis." *Studies in History and Philosophy of Biological and Biomedical Sciences*, 48, Part B: 162–74.

———. 2014b. "The Return of the Organism as Fundamental Explanatory Concept in Biology." *Philosophy Compass* 9(5): 347–59.

Niehoff, Debra. 2005. *The Language of Life: How Cells Communicate in Health and Disease*. Washington, DC: Joseph Henry Press.

Noble, Denis. 2006. *The Music of Life: Biology Beyond the Genome*. Oxford: Oxford University Press.

Nurse, Sir Paul. 2003. "The Great Ideas of Biology." *Clinical Medicine* 3(6): 560–68.

Nyhart, Lynn K. 1995. *Biology Takes Form: Animal Morphology and the German Universities, 1800–1900*. Chicago: University of Chicago Press.

Nyhart, Lynn K., and Scott Lidgard. 2011. "Individuals at the Center of Biology: Rudolf Leuckart's Polymorphismus der Individuen and the Ongoing Narrative of Parts and Wholes." With an Annotated Translation. *Journal of the History of Biology* 44(3): 373–443.

Oken, Lorenz. 1805. *Die Zeugung*. Bamberg: Bei Joseph Anton Goebhardt.

O'Malley, Maureen A., Alexander Powell, Johnathan F. Davies, and Jane Calvert. 2007. "Knowledge-Making Distinctions in Synethetic Biology." *BioEssays* 30(1): 57–65.

Otis, Laura. 1999. *Membranes: Metaphors of Invasion in Nineteenth-Century Literature, Science, and Politics*. Baltimore: Johns Hopkins University Press.

———. 2001. *Networking: Communicating with Bodies and Machines in the Nineteenth Century*. Ann Arbor: University of Michigan Press.

Pappas, Gus. 2005. "A New Literary Metaphor for the Genome or Proteome." *Biochemistry and Molecular Biology Education* 33(1): 15.

Parnes, Ohad. 2000. "The Envisioning of Cells." *Science in Context* 13(1): 71–92.

———. 2003. "From Agents to Cells: Theodore Schwann's Research Notes of the Years 1835–1838." In *Reworking the Bench: Research Notebooks in the History of Science*, edited by Frederick L. Holmes, Jürgen Renn, and Hans-Jörg Rheinberger, 119–40. Norwell, MA: Kluwer Academic Publishers.

Paton, R. C. 1992. "Towards a Metaphorical Biology." *Biology and Philosophy* 7: 279–94.

Pauly, Philip J. 1987. *Controlling Life: Jacques Loeb and the Engineering Ideal in Biology*. Oxford: Oxford University Press.

Pauwels, Eleonore. 2013. "Communication: Mind the Metaphor." *Nature* 500: 523–24.

Pawelec, Andrzej. 2006. "The Death of Metaphor." *Studia Linguistica* 123: 117–21.

Pawson, Tony. 1995. "Protein Modules and Signalling Networks." *Nature* 373 (6515): 573–80.

———. 2008. "Thinking About How Living Things Work." Kyoto Prize Lecture. Accessed December 8, 2010. http://Pawsonlab.mshri.on.ca.

Pawson, Tony, and Rune Linding. 2008. "Network Medicine." *FEBS Letters* 582: 1266–70.

Pawson, Tony, and Tracy M. Saxton. 1999. "Signaling Networks – Do All Roads Lead to the Same Genes?" *Cell* 97: 675–78.

Pawson, Tony, and John D. Scott. 1997. "Signaling Through Scaffold, Anchoring, and Adaptor Proteins." *Science* 278(5346): 2075–80.

Peirce, Charles Sanders. 1992. *The Essential Peirce: Selected Philosophical Writings, Volume 1 (1867–1893)*, edited by Nathan Houser and Christian Kloesel. Bloomington and Indianapolis: Indiana University Press.

Pentimalli, Francesca. 2007. "Technology: (Truly) On and Off at the Flick of a Switch." *Nature Reviews Genetics* 8, 654 (September 2007) [doi: 10.1038/nrg2184]. Accessed June 6, 2015.

Pepper, Stephen. 1942. *World Hypotheses*. Berkeley: University of California Press.

Perbal, Bernard. 2003. "Communication Is the Key." *Cell Communication and Signaling* 1(1): 3.

Peters, R. A. 1937. "Proteins and Cell-Organization." In *Perspectives in Biochemistry: Thirty-One Essays Presented to Sir Frederick Gowland Hopkins by Past and Present Members of His Laboratory*, edited by Joseph Needham and David. E. Green, 36–44. London: Cambridge University Press.

Picken, Laurence. 1960. *The Organization of Cells and Other Organisms*. Oxford: Clarendon Press.

Pigliucci, Massimo, and Maarten Boudry. 2011. "Why Machine-Information Metaphors Are Bad for Science and Science Education." *Science & Education* 20(5–6): 453–71.

Pinker, Steven. 2007. *The Stuff of Thought: Language as a Window into Human Nature*. New York: Viking.

Policard, A. 1964. *Cellules Vivantes et Populations Cellulaires: Dynamique et Structure Moléculaire*. Paris: Masson et Cie.

Psillos, Stathis. 1999. *Scientific Realism: How Science Tracks Truth*. London: Routledge.

Puck, Theodore T. 1972. *The Mammalian Cell as a Microorganism: Genetic and Biochemical Studies in Vitro*. San Francisco: Holden-Day.

Putnam, Hilary. 1981. *Reason, Truth and History*. Cambridge: Cambridge University Press.

———. 1992. *Realism with a Human Face*. Edited by James Conant. Cambridge, MA.: Harvard University Press.

Queller, D. C., and J. E. Strassman. 2009. "Beyond Society: The Evolution of Organismality." *Philosophical Transactions of the Royal Society B* 364: 3143–55.

Quine, Willard van Orman. 1969. "Ontological Relativity." In *Ontological Relativity & Other Essays*. New York: Columbia University Press.

Racker, Efraim. 1965. *Mechanisms in Bioenergetics*. New York: Academic Press.

Radman, Zdravko, ed. 1995. *From a Metaphorical Point of View: A Multidisciplinary Approach to the Cognitive Content of Metaphor*. Berlin: Walter de Gruyter.

Raff, Martin C. 1992. "Social Controls on Cell Survival and Cell Death." *Nature* 356(6368): 397–400.

———. 1996. "Death Wish." *The Sciences* July/August: 36–40.

———. 1998. "Cell Suicide for Beginners." *Nature* 396: 119–22.

Raff, M. C., B. A. Barres, J. F. Burne, H. S. Coles, Y. Ishizaki, and M. D. Jacobson. 1993. "Programmed Cell Death and the Control of Cell Survival: Lessons from the Nervous System." *Science* 262(5135): 695–700.

———. 1994. "Programmed Cell Death and the Control of Cell Survival." *Philosophical Transactions of the Royal Society of London B* (345): 265–68.

Rajasethupathy, P., S. J. Vaytadden, U. Bhall. 2005. "Systems Modeling: A Pathway to Drug Discovery." *Current Opinion in Chemical Biology* 9: 400–406.

Raspail, François-Vincent. 1843. *Histoire Naturelle de la santé et de la maladie chez les végétaux et chez les animaux en général et en particulier chez l'homme*. 2 vols. Paris: A. Levavasseur.

Rather, Lelland, J. 1990. *A Commentary on the Medical Writings of Rudolf Virchow: Based on Schwalbe's Virchow-Bibliographie, 1843–1901*. San Francisco: Jeremy Norman Co.

Ratzke, Christoph, and Jeff Gore. 2015. "Shaping the Crowd: The Social Life of Cells." *Cell Systems* 1: 310–12.

Reichenbach, Hans. 1951. *The Rise of Scientific Philosophy*. Berkeley: University of California Press.

Remak, Robert. 1855. *Untersuchungen über die Entwickelung der Wirbelthiere*. Berlin: Reimer.

Reynolds, Andrew S. 2000. "Statistical Method and the Peircean Account of Truth." *Canadian Journal of Philosophy* 30(2): 287–314.

———. 2007a. "The Theory of the Cell State and the Question of Cell Autonomy in Nineteenth and Early Twentieth-Century Biology." *Science in Context* 20(1): 71–95.

———. 2007b. "The Cell's Journey: From Metaphorical to Literal Factory." *Endeavour* 31(2): 65–70.

———. 2008a. "Amoebae as Exemplary Cells: The Protean Nature of an Elementary Organism." *Journal of the History of Biology* 41: 307–37.

———. 2008b. "Ernst Haeckel and the Theory of the Cell State: Remarks on the History of a Bio-Political Metaphor." *History of Science* 46(2): 123–52.

———. 2010. "The Redoubtable Cell." *Studies in History and Philosophy of the Biological and Biomedical Sciences* 41(3, Special issue, "Historical and Philosophical Perspectives on Cell Biology"): 194–201.

————. 2014. "The Deaths of a Cell: How Language and Metaphor Influence the Science of Cell Death." *Studies in the History and Philosophy of Biological and Biomedical Sciences* 48 (B): 175–84.

————. 2017. "Discovering the Ties that Bind: Cell-Cell Communication and the Development of Cell Sociology." In *Biological Individuality: Integrating Scientific, Philosophical, and Historical Perspectives*, edited by Scott Lidgard and Lynn K. Nyhart, 109–28. Chicago: University of Chicago Press.

————. 2018. "In Search of Cell Architecture: General Cytology and Early Twentieth Century Conceptions of Cell Organization." In *Visions of Cell Biology: Reflections Inspired by Cowdry's "General Cytology,"* edited by Karl Matlin, Jane Maienschein, and Manfred Laubichler. Chicago: University of Chicago Press.

Reynolds, Andrew S., and Norbert Hülsmann. 2008. "Ernst Haeckel's Discovery of Magosphaera Planula: A Vestige of Metazoan Origins?" *History and Philosophy of the Life Sciences* 30: 339–86.

Rich, Alexander. 1963. "Polyribosomes." *Scientific American* 1(209): 44–53.

Richards, Ivor A. 1936. *The Philosophy of Rhetoric*. Oxford: Oxford University Press.

————. 1955. *Speculative Instruments*. Chicago: University of Chicago Press.

Richards, Robert J. 2008. *The Tragic Sense of Life: Ernst Haeckel and the Struggle over Evolutionary Biology*. Chicago: University of Chicago Press.

Richmond, M. L. 2000. "T. H. Huxley's Criticism of German Cell Theory: An Epigenetic and Physiological Interpretation of Cell Structure." *Journal of the History of Biology*, 33(2): 247–89.

Rietman, Edward A., John Z. Colt, and Jack A. Tuszynski. 2011. "Interactomes, Manufacturomes and Relational Biology: Analogies between Systems Biology and Manufacturing Systems." *Theoretical Biology and Medical Modeling* 8: 19.

Ritter, William E. 1919. *The Unity of the Organism, or the Organismal Conception of Life*. 2 vols. Boston: Gorham Press.

Robinson, C., A. Sali, and W. Baumeister. 2007. "The Molecular Sociology of the Cell." *Nature* 450: 973–82.

Rodbell, Martin. 1980. "The Role of Hormone Receptors and GTP-regulatory Proteins in Membrane Transduction." *Nature* 285(5751): 17–22.

————.1995. "Signal Transduction: Evolution of an Idea. Nobel Lecture." *Environmental Health Perspectives* 103(4): 338–45.

Rodriguez, Xavier de Donato, and Alfonso Arroyo-Santos. 2011. "The Function of Scientific Metaphors: An Example of the Creative Power of Metaphors in Biological Theories." In *The Paths of Creation: Creativity in Science and Art*, edited by Alfred Marcos and Sixto J. Castro, 81–96. Bern: Peter Lang AG.

Roosth, Sophia. 2017. *Synthetic: How Life Got Made*. Chicago: University of Chicago Press.

Rorty, Richard. 1979. *Philosophy and the Mirror of Nature*. Princeton, NJ: Princeton University Press.

————. 1991. "Unfamiliar Noises: Hesse and Davidson on Metaphor." In

Objectivity, Relativism, and Truth: Philosophical Papers Volume 1, 162–72. Cambridge: Cambridge University Press.

Rosenblueth, Arturo, and Norbert Wiener. 1945. "The Role of Models in Science." *Philosophy of Science* 12(4): 316–21.

Rosenfeld, Simon. 2013. "Are the Somatic Mutation and Tissue Organization Field Theories of Carcinogenesis Incompatible?" *Cancer Informatics* 12: 221–29.

Rothbart, Daniel. 1984. "The Semantics of Metaphor and the Structure of Science." *Philosophy of Science* 51: 595–615.

Ruder, Warren, Ting Lu, and James J. Collins. 2011. "Synthetic Biology Moving into the Clinic." *Science* 333(2 September): 1248–52.

Runke, Jennifer. 2005. "Metaphors in Biology: More than Heuristic Devices." Paper presented at the meeting of the International Society for the History, Philosophy, and Social Studies of Biology at the University of Guelph, Ontario, July 13–17.

———. 2007. "Towards a Pragmatic Theory of Metaphor in Biology." Paper presented at the meeting of the International Society for the History, Philosophy, and Social Studies of Biology at the University of Exeter, England, July 25–29.

———. 2008. "Towards an Adequate Theory of Scientific Metaphor." PhD diss., University of Calgary.

Ruse, Michael. 2000. "Metaphor in Evolutionary Biology." *Revue Internationale de Philosophie* 54(214): 593–619.

———. 2005. "Darwinism and Mechanism: Metaphor in Science." *Studies in History and Philosophy of Biological and Biomedical Sciences* 36: 285–302.

———. 2010. *Science and Spirituality: Making Room for Faith in the Age of Science*. Cambridge: Cambridge University Press.

Russell, E. S. 1930. *The Interpretation of Development and Heredity: A Study in Biological Method*. Oxford: Clarendon Press.

Sachs, Julius von. 1887. *Lectures on the Physiology of Plants*. Translated by H. Marshall Ward. Oxford: Clarendon Press.

———.1892. "Physiologische Notizen, II. Beiträge zur Zellentheorie. Energiden und Zellen." *Flora* 75: 57–67.

Salmon, Wesley C. 1984. *Scientific Explanation and the Causal Structure of the World*. Princeton, NJ: Princeton University Press.

———. 1989. "Four Decades of Scientific Explanation." In *Scientific Explanation*, edited by P. Kitcher and W. C. Salmon, 3–219. Minneapolis: University of Minnesota Press.

Sapp, Jan. 1987. *Beyond the Gene: Cytoplasmic Inheritance and the Struggle for Authority in Genetics*. Oxford: Oxford University Press.

———. 1994. *Evolution by Association: A History of Symbiosis*. Oxford: Oxford University Press.

Saunders, J. W. Jr. 1966. "Death in Embryonic Systems." *Science* 154(3749): 604–12.

Schickore, Jutta. 2007. *The Microscope and the Eye: A History of Reflections, 1740–1870*. Chicago: University of Chicago Press.

Schleiden, Matthias. [1838] 1847. *Contributions to Phytogenesis.* Translated by Henry Smith. London: Sydenham Society.

Schön, Donald A. 1993. "Generative Metaphor: A Perspective on Problem-Setting in Social Policy." In *Metaphor and Thought.* 2nd ed., edited by Andrew Ortony, 137–63. Cambridge: Cambridge University Press.

Schultze, Max. [1861] 1987. "Über Muskelkörpchen und das was man eine Zelle zu nennen habe." *Archiv Anatomische Physiologische wissenschaftlich Medicin,* 1–27: Reprinted in *Klassische Schriften zur Zellenlehre von Matthias Jacob Schleiden, Theodor Schwann, Max Schultze,* eingeleitet und bearbeitet von Ilse Jahn, 131–54. Leipzig: Akademische Verlagsgesellschaft, Geest & Portig K.-G.

Schwann, Theodor. [1839] 1847. *Microscopical Researches into the Accordance in the Structure and Growth of Animals and Plants.* Translated by Henry Smith. London: Sydenham Society.

Searle, John. 1999. *Mind, Language and Society: Doing Philosophy in the Real World.* London: Weidenfeld & Nicholson.

Sedgwick, Adam. 1895. "On the Inadequacy of the Cellular Theory of Development, and on the Early Development of Nerves, Particularly of the Third Nerve and of the Sympathetic in Elasmobranchii." *Quarterly Journal of Microscopical Science* 37: 87–101.

———. 1896. "Further Remarks on Cell Theory, with a Reply to Mr. Bourne." *Quarterly Journal of Microscopical Science* 38: 331–37.

Shearer, Creswell. 1906. "On the Existence of Cell Communications Between Blastomeres." *Proceedings of the Royal Society of London, Series B* 77 (520): 498–505.

Shull, C.A. 1922. "Respiration of Thermophiles." *Botanical Gazette* 73(5) May: 419.

Siebold, Carl von. [1848] 1853. "On Unicellular Plants and Animals." *Quarterly Journal of Microscopical Science.* 1: 111–21; 195–206.

Siekevitz, Philip. 1957. "Powerhouse of the Cell." *Scientific American* 197(1): 131–44.

Sinding, Christiane. 1996. "Literary Genres and the Construction of Knowledge in Biology: Semantic Shifts and Scientific Change." *Social Studies of Science* 26(1): 43–70.

Sismondo, Sergio. 1996. *Science Without Myth: On Constructions, Reality, and Social Knowledge.* Albany, NY: SUNY Press.

Skloot, Rebecca. 2010. *The Immortal Life of Henrietta Lacks.* New York: Crown.

Slater, Matthew. 2013. "Cell Types as Natural Kinds." *Biological Theory* 7(2): 170–79.

Smithers, D. W. 1962. "An Attack on Cytologism." *The Lancet* 1(7228): 493-99.

———. 1969. "No Cell is an Island." *British Medical Journal.* 3(5673): 778.

Sonnenschein, Carlos, and Ana Soto. 1999. *The Society of Cells: Cancer and Control of Cell Proliferation.* New York: Springer.

Soskice, Janet Martin, and Rom Harré. 1995. "Metaphor in Science." In *From a Metaphorical Point of View: A Multidisciplinary Approach to the Cogni-*

tive Content of Metaphor, edited by Zdravko Radman, 289–307. Berlin: Walter de Gruyter.

Soto, Ana, Carlos Sonnenschein, and P. A. Miquel. 2008. "On Physicalism and Downward Causation in Developmental and Cancer Biology." *Acta Biotheoretica* 56: 257–74.

Stanier, Roger Y., and Cornelius B. van Neil. 1962. "The Concept of a Bacterium." *Archiv für Mikrobiologie* 42: 17–35.

Steinhart, Eric Charles. 2001. *The Logic of Metaphor: Analogous Parts of Possible Worlds*. Dordrecht: Kluwer.

Stepan, Nancy Leys. 1986. "Race and Gender: The Role of Analogy in Science." *Isis* 77: 61–277.

Sternberg, Paul W., and H. Robert Horovitz. 1989. "The Combined Action of Two Intercellular Signaling Pathways Specifies Three Cell Fates During Vulval Induction in C. elegans." *Cell* 58: 679–93.

Stevens, Stanley Smith, and Hallowell Davis. 1938. *Hearing: Its Psychology and Physiology*. New York: Wiley and Sons.

Stoker, M. G. P. 1972. "The Leeuwenhoek Lecture, 1971: Tumour Viruses and the Sociology of Fibroblasts." *Proceedings of the Royal Society of London B* 181: 1–17.

Strick, James E. 2000. *Sparks of Life: Darwinism and the Victorian Debates over Spontaneous Generation*. Cambridge, MA: Harvard University Press.

Suderman, Ryan, and Eric J. Deeds. 2013. "Machines vs. Ensembles: Effective MAPK Signaling through Heterogeneous Sets of Protein Complexes." *PLoS Computational Biology* 9(10): e1003278. Accessed November 17, 2013. doi. 10: 1371./journal.pcbi.1003278.

Sugimoto, Yoshikazu, Makoto Noda, Hitoshi Kitayama, and Yoji Ikawa. 1988. "Possible Involvement of Two Signaling Pathways in Induction of Neuron-Associated Properties by v-Ha-ras Gene in PC12 Cells." *The Journal of Biological Chemistry* 263(24): 12102–08.

Sutherland, Earl, and G. Alan Robison. 1966. "The Role of Cyclic-3', 5'-AMP in Response to Catecholamines and other Hormones." *Pharmacological Reviews* 18(1): 145–61.

Svegliati, Silvia, Giusi Marrone, Antonio Pezone, Tatiana Spadoni, Antonella Grieco, Gianluca Moroncini, Domenico Grieco, et al. 2014. "Oxidative DNA Damage Induces the ATM-Mediated Transcriptional Suppression of the Wnt-Inhibitor WIF-1 in Systemic Sclerosis and Fibrosis." *Science Signaling* (2 September) 7(341). Accessed Sept. 3, 2014.

Temkin, Owsei. 1949. "Metaphors of Human Biology." In *Science and Civilization*, edited by Robert C. Stauffer, 169–94. Madison: University of Wisconsin Press.

Tepperman, Jay. 1988. "A View of the History of Biology from an Islet of Langerhans." In *Endocrinology: People and Ideas*, edited by S. M. McCann, 285–334. Bethesda, MD: American Physiological Society.

Tepperman, Jay, and Helen M. Tepperman. 1965. "Adaptive Hyperlipogenesis—Late 1964 Model." *Annals of the New York Academy of Sciences* 131(1): 404–11.

Thievessen, Ingo, Peter M. Thompson, Sylvain Berlemont, Karen M. Plevock,

Sergey V. Plotnikov, Alice Zemljic-Horpf, Robert S. Ross, Michael W. Davidson, Gaudenz Danuser, Sharon L. Campbell, and Clare M. Waterman. 2013. "Vinculin-Actin Interaction Couples Actin Retrograde Flow to Focal Adhesions, but Is Dispensable for Focal Adhesion Growth." *The Journal of Cell Biology* 202(1): 163–77.

Tompa, Peter. 2012. "Intrinsically Disordered Proteins: A 10-Year Recap." *Trends in Biochemical Sciences* 37(12): 509–16.

Toulmin, Stephen. 1960. *The Philosophy of Science: An Introduction.* New York: Harper.

Trosko, J. E. 2011. "The Gap Junction as a 'Biological Rosetta Stone': Implications of Evolution, Stem Cells to Homeostatic Regulation of Health and Disease in the Barker Hypothesis." *Journal of Cell Communication and Signaling* 5: 53–66.

Trout, J. D. 2002. "Scientific Explanation and the Sense of Understanding." *Philosophy of Science* 69(2): 212–33.

Turbayne, Colin Murray. 1970. *The Myth of Metaphor.* Rev. ed., with forewords by Morse Peckham and Foster Tait and an appendix by Rolf Eberle. Columbia, SC: University of South Carolina Press.

Unger, Franz. 1852. *Botanische Briefe.* Wein: Carl Gerhold & Sohn.

Urry, Lisa A., Michael L. Cain, Peter V. Minorsky, Steven A. Wasserman, and Jane B. Reece. 2016. *Campbell Biology.* 11th ed. New York: Pearson.

Valk, Evin, Rainis Venta, Mihkel Örd, Ilona Faustova, Mardo Kõivomägi, and Mart Loog. 2014. "Multistep Phosphorylation Systems: Tunable Components of Biological Signaling Circuits." *Molecular Biology of the Cell* 25(22): 3456–60.

Van der Weele, Cor. 1999. *Images of Development: Environmental Causes in Ontogeny.* Albany: SUNY Press.

Van Fraassen, Bas C. 1980. *The Scientific Image.* New York: Oxford University Press.

———. 2008. *Scientific Representation: Paradoxes of Perspective.* New York: Oxford University Press.

Van Rijn-van Tongeren, Geraldine W. 1997. *Metaphors in Medical Texts.* Amsterdam: Rodopi.

Van Roey, K., Toby J. Gibson, and N. E. Davey. 2012. "Motif Switches: Decision-Making in Cell Regulation." *Current Opinions in Structural Biology* 22(3): 378–85.

Van Steenburgh, E. W. 1965. "Metaphor." *Journal of Philosophy* 62(22): 678–88.

Virchow, Rudolf. 1855. "Cellular-Pathologie." *Archiv für pathologische Anatomie und Physiologie und für klinische Medicin,* 8: 3–39.

———. [1858a] 1958. "On the Mechanistic Interpretation of Life." In *Disease, Life, and Man: Selected Essays by Rudolf Virchow.* Translated and with an introduction by Lelland J. Rather, 102–19. Stanford: Stanford University Press.

———. 1858b. *Die Cellularpathologie in ihrer Begründung auf physiologische und pathologische Gewebelehre.* Berlin: August Hirschwald.

Waddington, C. H. 1957. *Strategy of the Genes.* New York: MacMillan.

Wang, Xinjiang. 2011. "p53 Regulation: Teamwork between RING domains of Mdm2 and MdmX." *Cell Cycle* 10(24): 4225–29.

Weigmann, Katrin. 2004. "The Code, the Text and the Language of God." *EMBO Reports* 5(2): 116–18.

Weinberg, Robert A. 1998. *One Renegade Cell: How Cancer Begins.* New York: Basic Books.

———. 2014. "Coming Full Circle—From Endless Complexity to Simplicity and Back Again." *Cell* 157: 267–71.

Weindling, Paul. 1981. "Theories of the Cell State in Imperial Germany." In *Biology, Medicine and Society 1840–1940*, edited by Charles Webster. Cambridge: Cambridge University Press.

Weiss, Paul. 1947. "The Problem of Specificity in Growth and Development." *Yale Journal of Biology and Medicine* 19(3): 235–78.

Welch, G. Rickey. 1987. "The Living Cell as an Ecosystem: Hierarchical Analogy and Symmetry." *TREE* 2(10): 305–09.

Welch, G. Rickey, and James S. Clegg. 2010. "From Protoplasmic Theory to Cellular Systems Biology: A 150-Year Reflection." *American Journal of Physiology-Cell Physiology* 298(6):C1280–90.

———. 2012. "Cell Versus Protoplasm: Revisionist History." *Cell Biology International* 36: 643–47.

Whitehead, Alfred North. 1955. *The Concept of Nature (The Tarner Lectures Delivered in Trinity College, November 1919).* Cambridge: Cambridge University Press.

Whitman, C. O. 1893. "The Inadequacy of the Cell-Theory of Development." *Journal of Morphology* 8: 639–58.

Wiener, Norbert. 1961. *Cybernetics or Control and Communication in the Animal and the Machine.* 2nd ed. Cambridge, MA: MIT Press.

———. 1964. *God & Golem, Inc. A Comment on Certain Points Where Cybernetics Impinges on Religion.* Cambridge, MA: MIT Press.

Wilson, Edmund Beecher. 1896. *The Cell in Development and Inheritance.* New York: MacMillan Company.

———. 1900. *The Cell in Development and Inheritance*, 2nd revised and enlarged ed. New York: MacMillan.

———. 1925. *The Cell in Development and Heredity*, 3rd revised and enlarged ed. New York: MacMillan.

Wilson, Henry Van Peters. 1907. "Coalescence and Regeneration in Sponges." *Journal of Experimental Zoology* 5(2): 245–53.

Wilson, J. W. 1944. "Cellular Tissue and the Dawn of the Cell Theory." *Isis* 35: 168–75.

Winans, Stephen C. 2002. "Bacterial Esperanto." *Nature Structural Biology* 9(2): 83–84.

Winther, Rasmus Grønfeldt. Forthcoming. *When Maps Become the World: Abstraction and Analogy in Philosophy of Science.* Chicago: University of Chicago Press.

Wittgenstein, Ludwig. 1958. *Philosophical Investigations. The English Text of the Third Edition.* Translated by G. E. M. Anscombe. New York: MacMillan Publishing Co, Inc.

Woese, Carl. 2004. "A New Biology for a New Century." *Microbiology and Molecular Biology Reviews* 68(2): 173–86.

Wolpert, Lewis. [1991] 2008. *The Triumph of the Embryo*. Mineola, NY: Dover.

———. 2009. *How We Live and How We Die: The Secret Lives of Cells*. London: Faber and Faber.

Woodger, J. H. [1929] 1967. *Biological Principles: A Critical Study*. London: Routledge & Kegan Paul.

Woodward, J. 2011. "Scientific Explanation." In *The Stanford Encyclopedia of Philosophy*. Winter ed., edited by E. N. Zalta. Accessed December 12, 2011. http://plato.stanford.edu/archives/win2011/entries/scientific-explanation/.

Xiong, Wen, and James E. Ferrell Jr. 2003. "A Positive-Feedback-Based Bistable 'Memory Module' That Governs a Cell Fate Decision." *Nature* 426: 460–65.

Young, Robert M. 1990. "Darwinism and the Division of Labour." *Science as Culture* 9: 110–24.

Zahm, J. M., S. Hazgui, M. Matos, A. Ben Seddik, R. B. Nawrocky, M. Polette, P. Birembaut, and N. Bonnet. 2007. "Quantitative Videomicroscopic Analysis of the Sociologic Behavior of Non-Invasive and Invasive Tumor Cell Lines." *Cellular and Molecular Biology* 52(6): 54–60.

Zilsel, Edgar. 1942. "The Genesis of the Concept of Physical Law." *The Philosophical Review* 51(3): 245–79.

Zinder, Norton D. 1992. "Forty Years Ago: The Discovery of Bacterial Transduction." *Genetics* 132: 291–94.

Zinder, Norton D., and J. Lederberg. 1952. "Genetic Exchange in *Salmonella*." *Journal of Bacteriology* 64: 679–99.

Index